AF789133

The Role of NRF2 Transcription Factor

Edited by José Antonio Morales-Gonzalez,
Eduardo Osiris Madrigal-Santillan,
Marvin A. Soriano-Ursúa
and Ángel Morales-González

Published in London, United Kingdom

The Role of NRF2 Transcription Factor
http://dx.doi.org/10.5772/intechopen.111702
Edited by José Antonio Morales-Gonzalez, Eduardo Osiris Madrigal-Santillan, Marvin A. Soriano-Ursúa and Ángel Morales-González

Contributors
Mai A. Samak, Marvin A. Soriano-Ursúa, Ana L. Ocampo-Néstor, Melvin N. Rosalez, Eunice D. Farfán-García, E. Osiris Madrigal-Santillán, Ángel Morales-González, José A. Morales-González, Danyelly Bruneska Gondim Martins, Thaysa Walleria Aragão Santos, Maria Helena Menezes Estevam Alves, Rosângela Ferreira Frade de Araújo, Danyelly Bruneska Gondim Martins, Ananda Cristina de Aguiar, Francielle Maria de Araújo Barbosa, Glauber Moreira Leitão, Hongmin Wu, Xiancai Zhong, Neeraj Verma

© The Editor(s) and the Author(s) 2024
The rights of the editor(s) and the author(s) have been asserted in accordance with the Copyright, Designs and Patents Act 1988. All rights to the book as a whole are reserved by INTECHOPEN LIMITED. The book as a whole (compilation) cannot be reproduced, distributed or used for commercial or non-commercial purposes without INTECHOPEN LIMITED's written permission. Enquiries concerning the use of the book should be directed to INTECHOPEN LIMITED rights and permissions department (permissions@intechopen.com).
Violations are liable to prosecution under the governing Copyright Law.

Individual chapters of this publication are distributed under the terms of the Creative Commons Attribution 3.0 Unported License which permits commercial use, distribution and reproduction of the individual chapters, provided the original author(s) and source publication are appropriately acknowledged. If so indicated, certain images may not be included under the Creative Commons license. In such cases users will need to obtain permission from the license holder to reproduce the material. More details and guidelines concerning content reuse and adaptation can be found at http://www.intechopen.com/copyright-policy.html.

Notice
Statements and opinions expressed in the chapters are these of the individual contributors and not necessarily those of the editors or publisher. No responsibility is accepted for the accuracy of information contained in the published chapters. The publisher assumes no responsibility for any damage or injury to persons or property arising out of the use of any materials, instructions, methods or ideas contained in the book.

First published in London, United Kingdom, 2024 by IntechOpen
IntechOpen is the global imprint of INTECHOPEN LIMITED, registered in England and Wales, registration number: 11086078, 167-169 Great Portland Street, London, W1W 5PF, United Kingdom

British Library Cataloguing-in-Publication Data
A catalogue record for this book is available from the British Library

Additional hard and PDF copies can be obtained from orders@intechopen.com

The Role of NRF2 Transcription Factor
Edited by José Antonio Morales-Gonzalez, Eduardo Osiris Madrigal-Santillan, Marvin A. Soriano-Ursúa and Ángel Morales-González
p. cm.

This title is part of the Biochemistry Book Series, Volume 57
Topic: Cell and Molecular Biology
Series Editor: Andrei Surguchov
Topic Editor: Mary C. Maj
Associate Topic Editor: Felicia Ikolo

Print ISBN 978-0-85466-992-9
Online ISBN 978-0-85466-991-2
eBook (PDF) ISBN 978-0-85466-993-6
ISSN 2632-0983

We are IntechOpen,
the world's leading publisher of
Open Access books
Built by scientists, for scientists

7,300+
Open access books available

192,000+
International authors and editors

210M+
Downloads

156
Countries delivered to

Our authors are among the
Top 1%
most cited scientists

12.2%
Contributors from top 500 universities

CLARIVATE ANALYTICS BOOK CITATION INDEX INDEXED
WEB OF SCIENCE™

Selection of our books indexed in the Book Citation Index
in Web of Science™ Core Collection (BKCI)

Interested in publishing with us?
Contact book.department@intechopen.com

Numbers displayed above are based on latest data collected.
For more information visit www.intechopen.com

IntechOpen Book Series
Biochemistry
Volume 57

Aims and Scope of the Series

Biochemistry, the study of chemical transformations occurring within living organisms, impacts all of the life sciences, from molecular crystallography and genetics, to ecology, medicine and population biology. Biochemistry studies macromolecules - proteins, nucleic acids, carbohydrates and lipids –their building blocks, structures, functions and interactions. Much of biochemistry is devoted to enzymes, proteins that catalyze chemical reactions, enzyme structures, mechanisms of action and their roles within cells. Biochemistry also studies small signaling molecules, coenzymes, inhibitors, vitamins and hormones, which play roles in the life process. Biochemical experimentation, besides coopting the methods of classical chemistry, e.g., chromatography, adopted new techniques, e.g., X-ray diffraction, electron microscopy, NMR, radioisotopes, and developed sophisticated microbial genetic tools, e.g., auxotroph mutants and their revertants, fermentation, etc. More recently, biochemistry embraced the 'big data' omics systems. Initial biochemical studies have been exclusively analytic: dissecting, purifying and examining individual components of a biological system; in exemplary words of Efraim Racker, (1913 –1991) "Don't waste clean thinking on dirty enzymes." Today, however, biochemistry is becoming more agglomerative and comprehensive, setting out to integrate and describe fully a particular biological system. The 'big data' metabolomics can define the complement of small molecules, e.g., in a soil or biofilm sample; proteomics can distinguish all the proteins comprising e.g., serum; metagenomics can identify all the genes in a complex environment e.g., the bovine rumen.
This Biochemistry Series will address both the current research on biomolecules, and the emerging trends with great promise.

Meet the Series Editor

Andrei Surguchov, Ph.D., joined Baylor College of Medicine, Houston, TX, as a faculty member in 1992, where he studied the mechanisms of the genetic control of lipid metabolism. At the University of Utah, his research interests focused on cloning new genes encoding retinal proteins. He studied molecular and cellular mechanisms of neurodegenerative diseases and retinal degeneration at Washington University, St. Louis. Currently, his research focuses on the structure-function relationship of proteins involved in neurodegeneration and ocular diseases. Andrei Surguchov is an Editor-in-Chief at Biochemistry Research International and Associate Editor in several biomedical journals.

Meet the Volume Editors

José Antonio Morales-González graduated as a surgeon physician from FES-Iztacala, National Autonomous University of Mexico (UNAM). He engaged in doctoral studies in biological sciences at the UNAM. Dr. Morales-González has been awarded diverse recognitions: the Gustavo Baz Prada Medal by the UNAM and the Alfonso Caso Medal for academic merit by the UNAM and was distinguished by the National System of Researchers (SNI) as National Researcher Level 2 (2017–2025). He is the author of 103 internationally published articles, receiving more than 5,000 citations to these publications. He is also the editor and coordinator of 28 specialized books. Dr. Morales-González is a full-time tenured professor-researcher at the Escuela Superior de Medicina, IPN.

Marvin A. Soriano-Ursúa conducted his Ph.D. research in medicine at the Instituto Politécnico Nacional (IPN) of Mexico in 2012. Currently, he is working as a professor of physiology and medicinal chemistry and as a researcher in the Escuela Superior de Medicina, IPN. His research has included the rational design of new boron-containing compounds, but also compounds from natural sources, as potential new drugs. On the basis of this research, he has received several awards and honors, such as The National Youth Award for Science and Technology 2013 and the Boron in the Americas Award 2022 for exceptional service to boron research. He has authored more than 85 research articles, most about theoretical and experimental evaluation of bioactive natural and synthetic compounds.

Eduardo Osiris Madrigal-Santillan studied chemistry, bacteriology, and parasitology at the National School of Biological Sciences (ENCB) of the National Polytechnic Institute (IPN). He completed master's and doctorate studies in chemical biological sciences with a specialty in toxicological genetics at the same institution (ENCB-IPN). Dr. Madrigal-Santillan was recognized by the National System of Researchers (SNI) belonging to the National Council of Humanities, Sciences and Technologies (CONAHCYT-Mexico) as National Researcher Level 2 (2017–2025). He has published 102 national and international articles. He is the editor and coordinator of 13 specialized books. He is currently guest editor of the Plants journal of the Swiss publishing MDPI. Dr. Madrigal-Santillan is a full-time professor-researcher at the School of Medicine (ESM-IPN).

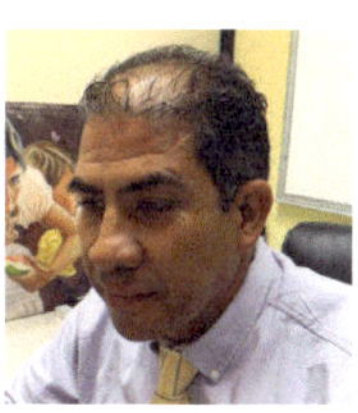

Ángel Morales-González engaged in doctoral studies specializing in systems engineering at SEPI-ESIME, IPN. He was distinguished by the National System of Researchers (SNI) as a National Researcher Candidate Level (2016–2018) and Level 1 (2019–2025). He is the author of 36 internationally and nationally published articles and editor and coordinator of six specialized books. At this moment, Dr. Morales-González is a full-time titular professor-researcher C at the ESCOM-Instituto Politécnico Nacional, México.

Contents

Preface

The present book, *The Role of NRF2 Transcription Factor*, represents the academic effort to describe the role of the transcription factor NRF2 in its functioning as a cytoprotector of the cell, tissue, organs, and the organism as a whole.

Its functions are varied as a regulator of endogenous antioxidant defenses, a detoxifier, and an anti-inflammatory agent, as well as contradictory when regulating anti-apoptotic and pro-apoptotic processes.

The book contains updated information on its regulation under normal conditions, theoretical studies on the NRF2 ligand-protein interaction, and its participation in cellular metabolism, neurological diseases, and immunological processes, as well as in cancer.

The participation of NRF2 in the cancer process can be considered the bad side of this transcription factor, which leads to different basic, translational, and clinical research studies to search for NRF2 activators or inhibitors to obtain a therapeutic window for these diseases. This can be considered of utmost importance. However finding a regulator of NRF2 is the priority in order to induce this factor to perform a specific function, for example, activating it with its pro-apoptotic function in cancer or activating it with its anti-apoptotic function in neurodegenerative diseases.

We hope that this book, *The Role of NRF2 Transcription Factor*, meets your expectations.

Dr. José Antonio Morales-González, Dr. Marvin A. Soriano-Ursúa, Dr. Eduardo Osiris Madrigal-Santillán and Dr. Ángel Morales-González
Superior School of Medicine,
National Polytechnic Institute,
Mexico City, Mexico

Chapter 1

Nrf2: The Guardian of Cellular Harmony – Unveiling Its Role in Cell Biology and Senescence

Mai A. Samak

Abstract

Nrf2, a key cellular regulator, plays a complex and multifaceted role in both protecting healthy cells and potentially promoting disease progression. This chapter delves into the intricate mechanisms by which Nrf2 exerts its protective effects, including combating carcinogens, maintaining cellular integrity, and inducing controlled cell death under severe stress. However, the chapter also explores the "dark side" of Nrf2, where its activity in cancer cells can contribute to chemoresistance, adaptation, and growth, hindering effective treatment. The chapter further investigates current research avenues for harnessing Nrf2's potential for therapeutic benefit. Strategies for both activation and inhibition are explored, highlighting the importance of context-dependent effects, balancing protection and potential harm, and minimizing off-target effects. Promising new directions, such as developing tissue-specific modulators, identifying predictive biomarkers, and combining Nrf2 modulators with other therapeutic approaches, are also discussed. By understanding Nrf2's complex and context-dependent roles, we can pave the way for the development of safer and more effective therapeutic strategies that leverage its beneficial effects while mitigating its potential drawbacks in various diseases.

Keywords: Nrf2, cellular homeostasis, Nrf2 and mitochondria crosstalk, cellular redox pathways, Nrf2 and cell senescence

1. Introduction

"In the intricate symphony of cellular biology, Nrf2 emerges as a central conductor orchestrating the delicate balance between cellular protection and senescence. This chapter delves into the multifaceted role of Nrf2, the guardian molecule, in maintaining cellular harmony. From its pivotal function in combating oxidative stress to its intricate involvement in regulating cellular senescence, the intricate dance of Nrf2 within the cellular landscape is unveiled, offering insights into its profound significance in both health and disease."

1.1 Overview of Nrf2 and its significance in cell biology

The transcription factor Nuclear factor-E2-related factor 2 (Nrf2) stands at the crossroads of cellular health, aging, and disease. Its role in maintaining redox balance, preventing senescence, and safeguarding against oxidative stress underscores its significance in cell biology [1]. Further research into Nrf2's regulatory mechanisms will pave the way for innovative therapies aimed at promoting healthy aging. Aging is an inevitable process that affects all living organisms.

1.1.1 Nrf2: a guardian of cellular redox balance

Under normal conditions, Nrf2 remains sequestered in the cytoplasm by its inhibitor, Keap1. However, when cells encounter oxidative stress or electrophilic insults, Nrf2 dissociates from Keap1 and translocates to the nucleus [2]. There, it binds to antioxidant response elements (AREs) in the promoter regions of target genes, activating their transcription.

1. *Target Genes*: Nrf2 induces the expression of various genes, including those encoding antioxidant enzymes and phase II detoxification enzymes.

2. *Cellular Senescence*: Premature senescence is associated with decreased Nrf2 levels, emphasizing its importance in aging processes [3].

1.1.2 Nrf2 and age-related diseases

The significance of Nrf2 extends beyond cellular senescence:

1. *Protection Against Oxidative Stress*: Nrf2 activation enhances the cellular antioxidant defense system implicated in various age-related diseases (e.g., neurodegenerative disorders, cardiovascular diseases, and cancer).

2. *Inflammation and Immunity*: Nrf2 also influences immune cell function, contributing to tissue repair and homeostasis [4, 5].

3. *Metabolism and Autophagy*: Dysregulation of Nrf2 can lead to metabolic imbalances and impaired autophagic processes [2, 3].

1.2 The battle against oxidative stress: exploring cellular defense mechanisms

Oxidative stress can lead to cellular damage, DNA mutations, and the development of various diseases, including cancer. However, living organisms have evolved intricate defense mechanisms to counteract the detrimental effects of oxidative stress [6]. Non-enzymatic antioxidants such as vitamins C and E, glutathione, Carotenoids, and Flavonoids neutralize ROS directly, preventing oxidative damage to biomolecules [5]. The synergistic action of these antioxidant systems helps to mitigate the harmful effects of oxidative stress and protect cellular components from damage.

1.2.1 DNA repair pathways

Oxidative stress can induce DNA damage, leading to mutations and genomic instability. To counteract this threat, cells employ a variety of DNA repair pathways to maintain genomic integrity. Base excision repair (BER), nucleotide excision repair (NER), and mismatch repair (MMR) pathways are activated in response to oxidative DNA damage, repairing lesions and preventing the accumulation of mutations [3, 7].

DOI: http://dx.doi.org/10.5772/intechopen.1005182

1.2.2 Cellular signaling pathways

Cellular signaling pathways play a crucial role in orchestrating the cellular response to oxidative stress. Moreover, the activation of stress-responsive kinases such as mitogen-activated protein kinases (MAPKs) and phosphoinositide 3-kinase (PI3K)/Akt pathway regulates cellular processes ensuring cell survival under adverse conditions [8].

2. Nrf2 pathway activation

When Nrf2 is activated, it enters the nucleus and turns on several hundred genes, known collectively as "survival genes." NRF2 activation decreases overall inflammation and oxidation of the body with a hormetic effect. Nutrients known to activate Nrf2 include sulforaphane from broccoli seed extract, curcumin from the turmeric plant, and carotenoids such as zeaxanthin and lutein.

Basal Regulation: Under normal physiological conditions, Keap1 targets Nrf2 for ubiquitination and subsequent degradation by the proteasome, maintaining low basal levels of Nrf2 activity [9].

Activation by Oxidative Stress: When cells encounter oxidative stress, this disrupts the interaction between Nrf2 and Keap1, leading to the stabilization and accumulation of Nrf2 in the cytoplasm.

Nuclear Translocation: It forms heterodimers with small Maf proteins and binds to antioxidant response elements (AREs) present in the promoter regions of target genes.

Transcriptional Activation: Binding of Nrf2 to AREs induces the transcription of a battery of cytoprotective genes encoding antioxidant enzymes and restores cellular redox homeostasis [9].

Pharmacological Activation: Pharmacological activation of the Nrf2 pathway has emerged as a potential therapeutic strategy for combating oxidative stress-related diseases. Nrf2 activators, including natural compounds (e.g., sulforaphane, curcumin) and synthetic compounds (e.g., bardoxolone methyl), enhance Nrf2 activity and promote cytoprotection [10, 11].

2.1 Nrf2 activation mechanisms

2.1.1 Keap1-mediated regulation

Under normal conditions, Nrf2 is sequestered in the cytoplasm by its inhibitor, Keap1, a substrate adaptor protein for a Cullin 3 (CUL3)-based E3 ubiquitin ligase [12]. This interaction promotes the ubiquitination and subsequent proteasomal degradation of NRF2, keeping its levels low under normal conditions. KEAP1 contains reactive cysteine residues that act as sensors for oxidative stress [13, 14]. When these cysteines are modified by reactive oxygen species (ROS) or electrophiles, KEAP1 undergoes a conformational change that prevents it from promoting the degradation of NRF2.

Post-Translational Modifications:

- Phosphorylation, acetylation, and ubiquitination modulate Nrf2 activity.
- These modifications affect Nrf2 stability, nuclear translocation, and DNA binding.

Coordinated Signaling Pathways [15, 16]*:*

- Nrf2 interacts with multiple signaling pathways, including PI3K/Akt, MAPK, and mTOR.
- These interactions fine-tune Nrf2 activity and influence cellular fate.

3. Nrf2 and antioxidant response

3.1 Role of Nrf2 in regulating antioxidant genes

Defective Nrf2 activation or downregulation of its target genes may compromise cellular antioxidant capacity and render cells more susceptible to oxidative damage and inflammatory insults [17].

Antioxidant genes regulated by Nrf2 [18, 19]

Antioxidant genes regulated by Nrf2 encompass a wide range of cytoprotective enzymes and proteins, including but not limited:

- Phase II Detoxification Enzymes: These enzymes facilitate the conjugation and elimination of electrophilic compounds and xenobiotics, thereby reducing cellular exposure to potentially harmful agents.
- Antioxidant Enzymes & Redox Regulators: Notably, genes encoding thioredoxin (TXN), thioredoxin reductase (TXNRD), and glutaredoxin (GLRX) are among the targets of Nrf2-mediated transcriptional regulation. These proteins play crucial roles in modulating cellular redox state and redox-sensitive signaling cascades.
- Stress Response Proteins: In addition to classical antioxidant enzymes, Nrf2 regulates the expression of stress response proteins that confer cellular resilience against diverse insults as heat shock proteins (HSPs).

3.2 Impact on cellular homeostasis

- *Regulation of Iron Homeostasis:* NRF2 is a potential modulator for orchestrating iron homeostasis and redox balance in cancer cells [20].
- *Dysfunction and Disease:* Studies in genetically modified mice demonstrate that increased oxidative stress due to NRF2 deficiency impairs endothelial function, reducing functional hyperemia in the brain [21].
- *Regulation of Stem Cell State and Function:* While NRF2 activation by tert-butylhydroquinone (t-BHQ) or sulforaphane (SFN) can suppress differentiation, it displays cell type-specific and/or stage-dependent impact on stem cell biology in response to various environmental cues [13, 21].

For adult stem cells (ASCs), NRF2 helps maintain their self-renewal, quiescence, and regenerative capacity while protecting against ASC depletion in response to stress and aging. NRF2 controls the proliferation of self-renewing embryonic stem cells (ESCs), three germ layer differentiation and cellular reprogramming NRF2-based molecular programs may govern stem cell state and function with age, and implications of this for age-related pathologies [22].

- *Stress Response:* NRF2 serves as a key mediator of the cellular stress response, coordinating adaptive responses to various stressors, including oxidative stress, electrophilic stress, and proteotoxic stress [23, 24].

4. Nrf2 in detoxification processes

4.1 Nrf2: orchestrator of cellular defenses through phase II enzyme induction

Nrf2 sits at the helm of the detoxification pathway, acting as a master transcription factor for Antioxidant Response Elements (AREs) to initiate the transcription and subsequent production of these detoxifying enzymes, mounting a coordinated cellular response to the encountered stress [25].

Phase II enzyme arsenal: neutralizing the threats [26]

Phase II enzymes comprise a diverse array of biochemical warriors, each specializing in neutralizing specific toxins and harmful molecules. Key players include:

4.1.1 Glutathione S-transferases (GSTs)

Mechanism:

- The "tagging" process with GSH involves a nucleophilic attack by the thiol group of GSH on the electrophilic center of the toxin. This forms a conjugate, making the toxin less reactive and more water-soluble.

  o *GSTM1:* Detoxifies polycyclic aromatic hydrocarbons (PAHs) found in cigarette smoke and pollution.

  o *GSTT1:* Protects against aflatoxins, carcinogens present in some food products.

  o *GSTP1:* Plays a role in metabolizing various drugs and pesticides.

Diversity:

- Genetic variations like single nucleotide polymorphisms (SNPs) can influence individual GST activity and detoxification capacity [27].

Induction:

- Nrf2 binding to AREs in GST gene promoters leads to increased mRNA and protein expression.

- Dietary factors like cruciferous vegetables can also boost GST activity through Nrf2 activation.

4.1.2 NAD(P)H: quinone oxidoreductase 1 (NQO1)

Mechanism:

- NQO1 catalyzes a two-step reduction process using NAD(P)H as a cofactor.
- First, it converts the quinone to a semiquinone radical.
- Second, it further reduces the semiquinone to a less harmful hydroquinone.

Induction [28]:

- Nrf2 is a potent activator of NQO1 expression, offering potential therapeutic strategies in cancer prevention and treatment.
- Dietary compounds like curcumin and sulforaphane can activate NQO1 through Nrf2 pathways.

4.1.3 UDP-glucuronosyltransferases (UGTs)

Mechanism:

- UGTs use UDP-glucuronic acid as a donor molecule to attach glucuronic acid to various compounds through an O-glycosidic bond.
- This conjugation process increases the molecule's size and polarity, rendering it inactive and water-soluble, facilitating renal or biliary excretion.

Individual Variability:

- Several genetic polymorphisms exist in UGT genes, leading to individual differences in enzyme activity.
- For example, the UGT2B7*2 allele reduces enzyme activity, potentially impacting drug response and contributing to adverse effects.

4.1.4 Heme oxygenase-1 (HO-1)

Mechanism:

- HO-1 catalyzes the degradation of heme, the iron-containing prosthetic group of hemoglobin, into Biliverdin, Iron, and Carbon monoxide
- By reducing iron release and generating beneficial metabolites, HO-1 protects against various inflammatory diseases like atherosclerosis, neurodegenerative disorders, and autoimmune diseases [29, 30].

4.2 NRF2: nature's guardian against xenobiotic and electrophilic threats

4.2.1 Xenobiotics

Nature of the Threats [31]*:*

- *Definition:* Foreign chemicals not naturally produced by or intended for biological organisms.
- *Sources:*
  - o Environmental pollutants (e.g., pesticides, air pollutants)
  - o Dietary toxins (e.g., mycotoxins, aflatoxins)
  - o Pharmaceuticals and medications

Mechanisms of Toxicity [32]*:*

- *Direct Damage:* Interfere with essential cellular processes or disrupt macromolecular structures (e.g., DNA, proteins).
- *Oxidative Stress:* Generate free radicals, leading to cellular damage and inflammation.
- *Epigenetic Alterations:* Can modify gene expression patterns, potentially contributing to disease development.

4.2.2 Electrophilic stressors

Definition: Molecules possessing an electron-deficient center, readily attracting electrons from other molecules.
Mechanisms of Stress [33]*:*

- *Covalent Binding:* React with cellular proteins, DNA, and lipids, causing functional disruptions and potential mutations.
- *Oxidative Stress:* Can trigger the generation of free radicals, leading to cellular damage and inflammation.
- *Disruption of Cellular Signaling:* Interfere with essential cell communication pathways.

Interplay of Xenobiotics and Electrophilic Stressors [34, 35]*:*

- Xenobiotics can be metabolized into electrophilic intermediates, amplifying their toxic effects.

- Electrophilic stressors can potentiate the toxicity of xenobiotics by increasing their reactivity or promoting free radical formation.

5. Nrf2 and cellular metabolism

Understanding the interactions between Nrf2 and cellular metabolism provides insights into the pathophysiology of metabolic disorders and offers opportunities for therapeutic interventions aimed at restoring metabolic homeostasis and improving health outcomes.

5.1 Influence of Nrf2 on metabolic pathways

- *Fatty Acid Metabolism:* Nrf2 regulates fatty acid metabolism by modulating the expression of genes involved in fatty acid oxidation (FAO) such as carnitine palmitoyltransferase 1 (CPT1) and acyl-CoA oxidase (ACOX), leading to increased utilization of fatty acids as an energy source [36, 37].

- *Glucose Homeostasis:* Activation of Nrf2 enhances glucose uptake and utilization by upregulating the expression of glucose transporters (e.g., GLUT1, GLUT3) and glycolytic enzymes (e.g., hexokinase, phosphofructokinase) [38].

- *Redox Regulation of Metabolism:* Nrf2 influences cellular metabolism through its antioxidant and redox regulatory functions. By enhancing antioxidant defense mechanisms, Nrf2 protects cells from oxidative damage and maintains redox balance, which is essential for proper metabolic function [36, 39].

- *Metabolic Diseases:* Dysregulation of Nrf2 signaling has been implicated in the pathogenesis of metabolic diseases, including obesity, type 2 diabetes, and metabolic syndrome. Impaired Nrf2 activation disrupts metabolic homeostasis, leading to metabolic dysfunction, insulin resistance, and inflammation [21, 36].

5.2 Influence of Nrf2 on mitochondrial function

Nrf2 plays a pivotal role in protecting mitochondria from oxidative stress and promoting their optimal function through multiple mechanisms:

5.2.1 Antioxidant response

Nrf2 activation upregulates antioxidant genes like superoxide dismutase (SOD) and catalase, neutralizing ROS and protecting mitochondrial membranes.

5.2.2 Mitochondrial biogenesis

Nrf2, the cellular defense chief, plays a crucial role in orchestrating this vital process, ensuring proper mitochondrial function and overall health [40]:

- *Transcriptional Activation:* Nrf2 binds to specific DNA sequences called Antioxidant Response Elements (AREs) in genes involved in mitochondrial biogenesis, such as PGC-1α. This activation leads to increased mRNA production for these crucial proteins.

- *Mitochondrial DNA Replication:* Nrf2 indirectly promotes mitochondrial DNA replication by upregulating genes involved in this process, ensuring new mitochondria have their own genetic machinery.

- *Mitochondrial Import:* Nrf2 regulates the import of proteins and other components necessary for mitochondrial assembly from the cytoplasm, ensuring proper construction [41].

Interplay with Other Pathways:

Nrf2 does not work in isolation. It interacts with other cellular signaling pathways to fine-tune biogenesis:

- *AMPK:* This energy sensor can activate Nrf2, promoting biogenesis in response to increased energy demands.

- *Sirtuins:* These longevity proteins can activate Nrf2, contributing to biogenesis and overall mitochondrial health [42].

5.2.3 Mitophagy

Mitophagy is a vital cellular process involved in the removal and recycling of damaged or dysfunctional mitochondria. Mitophagy plays a crucial role in maintaining mitochondrial quality control by eliminating damaged or dysfunctional mitochondria. Dysregulation of mitophagy has been implicated in various diseases, including neurodegenerative disorders, metabolic diseases, and aging, highlighting its importance in maintaining cellular homeostasis [43].

Nrf2: The Maestro of Mitophagy:

Nrf2 plays a critical role in orchestrating mitophagy through several mechanisms [44]:

- *Transcriptional Activation:* Nrf2 binds to AREs in genes involved in mitophagy pathways, such as Parkin, PINK1, and BNIP3, leading to increased production of these key proteins.

- *Regulation of Post-Translational Modifications:* Nrf2 can influence the phosphorylation and ubiquitination of proteins involved in mitophagy, facilitating their interaction and efficient targeting of damaged mitochondria.

- *Mitochondrial Quality Control:* Nrf2 promotes the expression of genes that monitor mitochondrial health, allowing for early detection of damage and triggering mitophagy when necessary.

5.2.4 Calcium homeostasis

Nrf2 emerges as a crucial player in maintaining calcium homeostasis, influencing intracellular and extracellular calcium levels and impacting various physiological processes [45]:

- *Regulation of Calcium Channels:* Nrf2 can modulate the expression and activity of calcium channels in various cell types, influencing calcium influx and intracellular levels [45].

- *Endoplasmic Reticulum (ER) Stress Response:* Nrf2 activation contributes to the unfolded protein response (UPR) in the ER, a stress response pathway that can influence calcium signaling and homeostasis [46].

- *Mitochondrial Calcium Handling:* Nrf2 indirectly impacts mitochondrial calcium uptake and release, potentially influencing energy production and cell function [43].

Impact on Health and Disease:
Understanding Nrf2's role in calcium homeostasis sheds light on various health aspects:

- *Neurodegenerative Diseases:* Dysregulation of calcium signaling and Nrf2 dysfunction are implicated in Alzheimer's and Parkinson's diseases.

- *Cardiovascular Diseases:* Nrf2 may influence calcium handling in heart muscle cells, potentially impacting cardiac function and arrhythmia risk.

- *Bone Health:* Nrf2 might indirectly affect bone formation and resorption through its influence on calcium homeostasis and oxidative stress response [47].

6. Nrf2 in cell proliferation and differentiation

Nrf2, the master regulator of cellular defense, plays a complex and multifaceted role in both cell proliferation and differentiation.

6.1 Nrf2's involvement in cell growth and development

Promoting Cell Growth [48]*:*

- *Antioxidant and Anti-inflammatory Effects:* By countering oxidative stress and inflammation, Nrf2 creates a favorable environment for cell growth, especially in tissues with high renewal rates like skin and intestine.

- *Metabolic Regulation:* Nrf2 can influence energy metabolism and nutrient availability, indirectly impacting cell growth under certain conditions.

- *Interaction with Growth Factors:* Nrf2 may interact with specific growth factor signaling pathways, potentially promoting cell proliferation in response to growth signals.

Modulating Cell Differentiation:

- *Transcriptional Regulation:* Nrf2 can regulate genes involved in differentiation pathways, either promoting or inhibiting it depending on the context.

- *Cell-Type Specificity:* The influence of Nrf2 on differentiation varies significantly between different cell types. For example, it can promote differentiation in embryonic stem cells but inhibit it in muscle cells.

- *Oxidative Stress Response:* Under high oxidative stress, Nrf2 activation may prioritize survival and stress response over differentiation, temporarily delaying the process [41].

6.2 Crosstalk between Nrf2 and other signaling pathways

Mechanisms of Crosstalk:

- *Direct Protein-Protein Interactions:* Nrf2 can physically interact with proteins in other pathways, influencing their activity and localization.

- *Transcriptional Regulation:* Nrf2 can modulate the expression of genes involved in other pathways, and vice versa.

- *Post-Translational Modifications:* Phosphorylation, ubiquitination, and other modifications can influence the interaction and activity of proteins in both pathways.

Overview of key signaling pathways that interact with Nrf2:

PI3K/Akt Pathway: The phosphoinositide 3-kinase (PI3K)/Akt pathway plays a crucial role in cell survival, proliferation, and metabolism. Activation of Akt can promote Nrf2 activation by phosphorylating Nrf2 or its upstream regulators, leading to increased Nrf2 stability and nuclear translocation. Conversely, Nrf2 activation can enhance Akt signaling by suppressing phosphatase and tensin homolog (PTEN) expression, thereby amplifying cell survival pathways in response to oxidative stress [49].

MAPK Pathway: The mitogen-activated protein kinase (MAPK) signaling pathway, including extracellular signal-regulated kinase (ERK), c-Jun N-terminal kinase (JNK), and p38 MAPK, regulates various cellular processes such as proliferation, differentiation, and stress responses. Crosstalk between MAPK and Nrf2 pathways occurs at multiple levels, with MAPKs influencing Nrf2 activity through phosphorylation-mediated modulation of Nrf2 stability, subcellular localization, and transcriptional activity. Conversely, Nrf2 activation can attenuate MAPK signaling by inhibiting inflammatory gene expression and ROS production [46, 50].

NF-κB Pathway: Crosstalk between Nrf2 and NF-κB pathways occurs bidirectionally, with Nrf2 activation inhibiting NF-κB signaling by suppressing pro-inflammatory gene expression and NF-κB activation. Conversely, NF-κB activation can repress Nrf2 activity by inducing the expression of Keap1 or promoting Nrf2 degradation, thus modulating the balance between pro-inflammatory and antioxidant responses [51].

HIF-1α Pathway: The hypoxia-inducible factor 1-alpha (HIF-1α) pathway regulates cellular responses to hypoxia and oxygen availability. Coordination between Nrf2 and HIF-1α pathways ensures cellular adaptation and survival under fluctuating oxygen and nutrient levels [52].

AMPK Pathway: The adenosine monophosphate-activated protein kinase activation can stimulate Nrf2 signaling by phosphorylating Nrf2 or its regulators, leading to increased Nrf2 activity and antioxidant gene expression. Conversely, Nrf2 activation can enhance AMPK signaling by promoting mitochondrial biogenesis and fatty acid oxidation, thus coordinating cellular responses to energy stress and oxidative stress [53].

7. Nrf2 and cell senescence

7.1 Unveiling the mysteries of Nrf2 and the aging cell

Nrf2, the master regulator of cellular defense, stands at the crossroads of aging and cellular health. Its influence on the aging process is intricate and multifaceted, offering both potential benefits and challenges.

The Double-Edged Sword:

- *Cancer Risk:* While Nrf2 protects healthy cells, its overactivation in cancer cells can promote their survival and growth. This highlights the need for a nuanced understanding of Nrf2's role in different contexts [54].

- *Hormonal Decline:* Estrogen and other hormones can activate Nrf2, and their decline with age might contribute to reduced Nrf2 activity and accelerated aging [55].

- *Cellular Senescence:* Nrf2 can both promote and inhibit cellular senescence, a state of cell cycle arrest associated with aging. Understanding the specific mechanisms involved is crucial [1].

7.1.1 Unraveling the mechanisms: Nrf2's fight against aging hallmarks

These mechanisms are interconnected, and Nrf2's influence on these hallmarks of aging is likely complex and context-dependent.

7.1.1.1 DNA damage

- *Antioxidant Response:* Nrf2 upregulates genes encoding antioxidant enzymes like superoxide dismutase (SOD) and catalase, which neutralize free radicals that can damage DNA [56].

- *DNA Repair Pathways:* Nrf2 activates genes involved in DNA repair mechanisms like base excision repair and non-homologous end joining, facilitating the repair of damaged DNA strands [57].

- *Modulation of Inflammatory Signaling:* Nrf2 suppresses chronic inflammation, which can indirectly contribute to DNA damage through the generation of reactive species [56].

7.1.1.2 Telomere shortening

- *Telomerase Regulation:* While the direct role of Nrf2 in telomerase activity remains under investigation [58], some evidence suggests that it might indirectly promote telomerase expression, protecting telomeres from shortening [59, 60].

- *DNA Repair Mechanisms:* As aforementioned, Nrf2's activation of DNA repair pathways can help maintain telomere integrity by repairing damage at the ends of chromosomes [57].

- *Antioxidant Defense:* By reducing oxidative stress, Nrf2 protects telomeres from free radical-induced damage, potentially slowing down their shortening [59, 61].

7.1.1.3 Protein misfolding

- *Chaperone Induction:* Nrf2 upregulates genes encoding molecular chaperones, proteins that help proteins fold correctly and prevent misfolding, a key contributor to protein aggregation and cellular dysfunction in aging [62].

- *Proteasomal Activity:* Nrf2 can enhance the activity of the proteasome, the cellular machinery responsible for degrading misfolded proteins, ensuring efficient protein turnover and preventing aggregation [63].

- *Autophagy Activation:* Nrf2 indirectly promotes autophagy, a cellular process that removes damaged proteins and organelles, further contributing to protein quality control [64].

7.1.2 The impact of aging on Nrf2 role in different organ systems

Nrf2 plays a crucial role in combating various aging hallmarks across different organ systems. However, its specific function and the impact of aging on its activity vary depending on the tissue:

Brain [65, 66]*:*

Aging is associated with increased oxidative stress and impaired antioxidant defense mechanisms in the brain, contributing to neurodegenerative diseases such as Alzheimer's disease and Parkinson's disease. Nrf2 plays a crucial role in protecting neurons from oxidative damage and promoting neuroprotection through the upregulation of antioxidant enzymes and phase II detoxification enzymes.

Liver:

The liver is a major site of metabolic and detoxification processes, making it susceptible to oxidative stress and age-related changes in Nrf2 signaling. Aging is associated with decreased Nrf2 activity and reduced capacity for detoxification and increased susceptibility to liver damage [67].

Kidneys:

Dysregulation of Nrf2 signaling in the kidneys has been implicated in the pathogenesis of age-related kidney diseases, including chronic kidney disease (CKD) and renal fibrosis [68].

Cardiovascular System:

Nrf2 plays a protective role in the cardiovascular system by upregulating antioxidant genes and inhibiting inflammatory pathways. However, aging-related decline in Nrf2 activity and expression may impair vascular function, promote endothelial dysfunction, and increase susceptibility to cardiovascular diseases [69].

Lungs:

The lungs are constantly exposed to environmental toxins and oxidative stress, making them vulnerable to age-related changes in Nrf2 signaling [70]. Aging-related decline in Nrf2 activity increases susceptibility to oxidative damage and respiratory diseases, such as chronic obstructive pulmonary disease (COPD) and pulmonary fibrosis [71].

Muscle:

Nrf2 helps maintain muscle function by promoting mitochondrial health and preventing protein misfolding. Aging can decrease Nrf2 activity in muscles, leading to sarcopenia (muscle loss) and decreased physical function [72].

Skin:

Nrf2 shields the skin from UV radiation and promotes wound healing. Aging reduces Nrf2 activity in the skin, contributing to wrinkles, age spots, and decreased wound healing capacity [73].

Immune System:

Nrf2 balances immune responses and prevents excessive inflammation. Aging can lead to dysregulation of Nrf2 in immune cells, contributing to autoimmune diseases and decreased immune function [74].

Bone:

Nrf2 regulates bone remodeling and protects against bone loss. Aging can decrease Nrf2 activity in bone cells, contributing to osteoporosis [75].

8. Nrf2 in Cancer biology

8.1 Dual role of Nrf2 in cancer as a protector and potential target

Nrf2, the cellular defense chief, plays a complex and paradoxical role in cancer. While it acts as a protector in healthy cells by shielding them from harmful stressors, it can emerge as a potential target in cancer cells, promoting their survival and resistance to therapy.

8.1.1 Nrf2's protective shield in healthy cells: A multifaceted defense

Nrf2 plays a crucial role in safeguarding healthy cells from various stressors and preventing cancer initiation.

Maintaining Cellular Integrity:

- *Inflammation Modulation:* Nrf2 suppresses chronic inflammation, which can contribute to cancer development by creating a pro-tumorigenic environment. Nrf2 does this by regulating the expression of inflammatory genes and promoting the resolution of inflammation [71].

- *Protein Quality Control:* Nrf2 activates genes encoding molecular chaperones, proteins that help proteins fold correctly and prevent misfolding. Misfolded proteins can accumulate and damage cells, potentially contributing to cancer development. Nrf2 also promotes proteasomal activity, the cellular machinery responsible for degrading damaged and misfolded proteins, ensuring efficient protein turnover and preventing protein aggregation [74].

- *Autophagy Activation:* Nrf2 indirectly promotes autophagy, a cellular process that removes damaged organelles and proteins. This helps maintain cellular health and prevents the accumulation of harmful components that could contribute to cancer development [60].

Cell Death Induction:

- *Under Severe Stress:* When cellular damage is beyond repair, Nrf2 can switch gears and trigger programmed cell death (apoptosis) in severely damaged cells. This prevents them from accumulating mutations and potentially becoming cancerous.

8.1.2 The dark side of Nrf2 in cancer cells

While Nrf2 acts as a guardian in healthy cells, its influence on cancer cells takes a sinister turn. The mechanisms by which Nrf2 becomes a potential target in cancer, promoting their survival and resistance to therapy [76, 77]:

Fueling Cancer Cell Survival:

- *Antioxidant Shield:* Nrf2 activates genes encoding antioxidant enzymes, originally meant to neutralize harmful free radicals. However, in cancer cells, these enzymes can also scavenge therapeutic drugs and reduce their effectiveness. For example, Nrf2 can induce enzymes that detoxify chemotherapeutic agents, rendering them less potent [78].

- *Metabolic Adaptation:* Nrf2 helps cancer cells adapt to stressful environments like nutrient deprivation or hypoxia (low oxygen). This allows them to survive and continue growing even under conditions that might otherwise kill them. Nrf2 achieves this by regulating genes involved in alternative metabolic pathways, enabling cancer cells to scavenge energy and survive in harsh environments [79].

- *DNA Repair Enhancement:* While Nrf2-mediated DNA repair is beneficial in healthy cells, it can become detrimental in cancer. Nrf2 can enhance DNA repair mechanisms in cancer cells, allowing them to repair damage caused by chemotherapy or radiation and continue proliferating. This contributes to tumor recurrence and treatment resistance [80].

Adaptation and Growth [81, 82]*:*

- *Proliferation Signals:* Nrf2 can interact with certain growth factor pathways in cancer cells, leading to increased proliferation and tumor growth. This highlights the context-dependent nature of Nrf2's role, where it can switch from protective to pro-tumorigenic depending on the specific cellular context.

- *Angiogenesis Promotion:* Nrf2 can induce genes involved in angiogenesis (formation of new blood vessels), providing cancer cells with a vital supply of nutrients and oxygen for their continued growth and expansion.

- *Metastasis Support:* Nrf2 might contribute to metastasis by promoting epithelial-mesenchymal transition (EMT), a process where cancer cells acquire migratory and invasive properties, enabling them to spread to other organs.

Therapeutic Strategies: Balancing the Act [83, 84]*:*

- *Harnessing Nrf2's Protective Potential:* In healthy tissues or early-stage cancers, low-dose Nrf2 activators might offer benefits by promoting DNA repair and preventing further carcinogenesis.

- *Targeting Nrf2 in Established Cancers:* Developing Nrf2 inhibitors or drugs that disrupt its interaction with specific cancer-promoting pathways is being explored to overcome chemoresistance and enhance treatment efficacy.

- *Personalized Medicine:* Understanding individual variations in Nrf2 activity and genetic makeup is crucial for tailoring therapeutic strategies to maximize benefits and minimize risks associated with targeting Nrf2 in cancer.

8.2 Overview of research on Nrf2 and chemoresistance

Research provided a rationale for targeting Nrf2 signaling as a therapeutic strategy to overcome chemoresistance and improve treatment outcomes in cancer patients [85].

8.2.1 Mechanisms of Nrf2-mediated chemoresistance

Nrf2-mediated metabolic reprogramming can also contribute to chemoresistance by altering cellular metabolism, promoting glycolysis, and enhancing cellular antioxidant capacity, thereby protecting cancer cells from chemotherapy-induced oxidative stress and apoptosis [86].

8.2.2 Clinical relevance and prognostic significance

Clinical studies have shown that patients with high Nrf2 expression or Nrf2 activation have lower response rates to chemotherapy, shorter progression-free survival, and reduced overall survival compared to patients with low Nrf2 expression [84].

8.2.3 Therapeutic targeting of Nrf2 signaling

Strategies to inhibit Nrf2 signaling have been explored as a means to overcome chemoresistance and enhance the efficacy of chemotherapy in cancer treatment.

Small molecule inhibitors of Nrf2 or its upstream regulators, such as Keap1 or glycogen synthase kinase 3β (GSK-3β), have been developed and investigated for their ability to sensitize cancer cells to chemotherapy and overcome drug resistance [87].

9. Therapeutic potential of Nrf2 modulation

Deeper scientific research into the molecular mechanisms of Nrf2 signaling and the development of Nrf2-targeted therapeutics may pave the way for novel treatments for oxidative stress-related diseases and contribute to improved health outcomes and quality of life.

9.1 Potential strategies for targeting Nrf2 in disease treatment

Nrf2's complex role in both protecting healthy cells and potentially promoting disease progression presents a unique challenge for therapeutic development.

Activation Strategies: Unlocking Potential Benefits

Activating Nrf2 holds promise for treating various diseases by leveraging its protective and reparative functions in cells.

Natural Compounds:

- *Dietary Sources:* Fruits and vegetables rich in phytochemicals like sulforaphane (broccoli, Brussels sprouts), curcumin (turmeric), resveratrol (grapes, red wine), and epigallocatechin gallate (green tea) can activate Nrf2 through various mechanisms.

- *Marine-Derived Compounds:* Certain marine compounds like fucoidan (found in brown algae) and astaxanthin (microalgae) show promise in activating Nrf2 and offering potential benefits in neurodegenerative diseases and age-related decline. Ongoing research is exploring their efficacy and safety profiles.

Nrf2 Activators:

- *Drug Development:* Some activators like bardoxolone methyl and dimethyl fumarate are undergoing clinical trials for specific diseases like chronic kidney disease and multiple sclerosis.

- *Specificity Considerations:* Designing Nrf2 activators with tissue-specific or disease-specific activity is a major focus to minimize potential off-target effects.

Gene Therapy:

- *Future Potential*: In rare cases with severe Nrf2 dysfunction due to genetic mutations, future gene therapy approaches might explore delivering functional Nrf2 genes directly into cells.

9.2 Discussion on current challenges and future directions

Challenges and considerations:

9.2.1 Context-dependent effects

- *Disease Specificity:* Nrf2's role can be vastly different depending on the disease. For instance, it might be protective in neurodegenerative diseases but contribute to chemoresistance in cancer.

- *Disease Stage:* Nrf2's role can even change within the same disease depending on its stage. In early-stage cancer, it might offer some protection, while in advanced stages, it might promote tumor growth.

- *Individual Genetic Variations:* Genetic variations in Nrf2 itself or its regulatory pathways can significantly influence its activity and response to modulators.

9.2.2 Balancing protection and harm

- *Preserving Nrf2's Protective Roles:* Inhibiting Nrf2 completely can disrupt its essential functions in healthy tissues, potentially increasing susceptibility to oxidative stress, inflammation, and other health problems.

- *Understanding Off-Target Effects:* Careful evaluation of potential off-target effects and long-term safety is essential before clinical application.

- *Developing Tissue-Specific Modulators:* Designing Nrf2 modulators with activity restricted to specific tissues or disease-relevant pathways holds promise for minimizing off-target effects and maximizing therapeutic benefits.

9.2.3 Off-target effects

- *Direct Nrf2 Inhibition:* Directly targeting Nrf2 protein can inadvertently affect its interactions with other proteins involved in various cellular processes, potentially leading to unforeseen side effects.

- *Upstream and Downstream Effects:* Targeting upstream regulators or downstream pathways of Nrf2 can also have unintended consequences due to the complex network of interactions within the cell.

- *Preclinical Testing and Personalized Medicine:* Rigorous preclinical testing and personalized medicine approaches that consider individual genetic variations and disease contexts can help mitigate off-target effects and identify patients who might benefit most from specific Nrf2 modulation strategies.

9.2.4 Combining Nrf2 modulators with other therapies

- *Synergistic Effects:* Combining Nrf2 modulators with other established therapies, such as chemotherapy or radiation for cancer, or immune checkpoint inhibitors, could offer synergistic effects and improve treatment outcomes.

- *Overcoming Resistance:* Combining Nrf2 modulators with strategies that target resistance mechanisms in diseases like cancer can help overcome current therapeutic challenges and improve patient survival.

- *Tailored Combinations:* Personalized combinations based on individual genetic profiles, disease stage, and other factors can optimize treatment efficacy and minimize potential side effects [1, 85].

Acknowledgements

The authors acknowledge the use of Grammarly and ProWritingAid tools for language polishing of the manuscript.

Author details

Mai A. Samak
Faculty of Medicine, Department of Medical Histology and Cell Biology, Zagazig University, Egypt

*Address all correspondence to: dr.maiamin@yahoo.com

IntechOpen

© 2024 The Author(s). Licensee IntechOpen. This chapter is distributed under the terms of the Creative Commons Attribution License (http://creativecommons.org/licenses/by/3.0), which permits unrestricted use, distribution, and reproduction in any medium, provided the original work is properly cited.

References

[1] Yuan H, Xu Y, Luo Y, Wang N-X, Xiao J-H. Role of Nrf2 in cell senescence regulation. Molecular and Cellular Biochemistry. 2021;**476**(1):247-2591

[2] Hayes JD, Dinkova-Kostova AT. NRF2: A multifaceted regulator of redox homoeostasis. Nature Reviews Molecular Cell Biology. 2014;**15**(7):509-521

[3] López-Otín C, Blasco MA, Partridge L, Serrano M, Kroemer G. The hallmarks of aging. Cell. 2013;**153**(6):1194-1217

[4] Kensler TW, Wakabayashi N, Biswal S. Cell survival responses to environmental stresses via the Keap1-Nrf2-ARE pathway. Annual Review of Pharmacology and Toxicology. 2007;**47**:89-116

[5] Cuadrado A, Rojo AI, Wells G, Hayes JD, Cousin SP, Rumsey WL, et al. Therapeutic targeting of the NRF2 and KEAP1 partnership in chronic diseases. Nature Reviews Drug Discovery. 2019;**18**(4):295-317

[6] Iqbal MJ, Kabeer A, Abbas Z, Siddiqui HA, Calina D, Sharifi-Rad J, et al. Interplay of oxidative stress, cellular communication, and signaling pathways in cancer. Cell Communication and Signaling. 2024;**22**:71

[7] Halliwell B, Gutteridge JM. Free Radicals in Biology and Medicine. USA: Oxford University Press; 2015

[8] Sies H. Oxidative stress: Oxidants and antioxidants. Experimental Physiology: Translation and Integration. 1997;**82**(2):291-295

[9] Finkel T, Holbrook NJ. Oxidants, oxidative stress and the biology of ageing. Nature. 2000;**408**(6809):239-247. This seminal review discusses the role of oxidative stress in aging and age-related diseases

[10] Zhang DD. Mechanistic insights into the Keap1-Nrf2-ARE pathway. Chemical Research in Toxicology. 2006;**18**(1):19-284

[11] Mukherjee AG, Gopalakrishnan AV. The mechanistic insights of the antioxidant Keap1-Nrf2 pathway in oncogenesis: A deadly scenario. Medical Oncology. 2023;**40**(248):3

[12] Padmanabhan B, Tong KI, Ohta T, Nakamura Y, Scharlock M, Ohtsuji M, et al. Structural basis for defects of Keap1 activity provoked by its point mutations in lung cancer. Molecular Cell. 2006;**21**(5):689-700

[13] Pouremamali F, Pouremamali A, Dadashpour M, Soozangar N, Jeddi F. An update of Nrf2 activators and inhibitors in cancer prevention/promotion. Cell Communication and Signaling. 2022;**20**(1):100

[14] Villeneuve NF, Lau A, Zhang DD. Regulation of the Nrf2–Keap1 antioxidant response by the ubiquitin proteasome system: An insight into cullin-ring ubiquitin ligases. Antioxidants & Redox Signaling. 2010;**13**(11):1699-1712

[15] Stewart D, Killeen E, Naquin R, Alam S, Alam J. Degradation of transcription factor Nrf2 via the ubiquitin-proteasome pathway and stabilization by cadmium. Journal of Biological Chemistry. 2003;**278**(4):2396-2402

[16] McMahon M, Itoh K, Yamamoto M, Hayes JD. Keap1-dependent proteasomal

degradation of transcription factor Nrf2 contributes to the negative regulation of antioxidant response element-driven gene expression. Journal of Biological Chemistry. 2003;**278**(24):21592-21600

[17] Park J-S, Rustamov N, Roh Y-S. The roles of NFR2-regulated oxidative stress and mitochondrial quality control in chronic liver diseases. Antioxidants. 2023;**12**(11):1928

[18] Zhang H, Davies KJ, Forman HJ. Oxidative stress response and Nrf2 signaling in aging. Free Radical Biology and Medicine. 2015;**88**:314-336

[19] Zhang C. Essential functions of iron-requiring proteins in DNA replication, repair and cell cycle control. Protein & Cell. 2017;**5**(10):750-760

[20] Tsai JJ, Dudakov JA, Takahashi K, Shieh JH, Velardi E, Holland AM, et al. Nrf2 regulates haematopoietic stem cell function. Nature Cell Biology. 2013;**15**(3):309-316

[21] Bellezza I, Giambanco I, Minelli A, Donato R. Nrf2-Keap1 signaling in oxidative and reductive stress. Biochimica et Biophysica Acta (BBA) – Molecular Cell Research. 2018;**1865**(5):721-733

[22] Jaramillo MC, Zhang DD. The emerging role of the Nrf2-Keap1 signaling pathway in cancer. Genes & Development. 2013;**27**(20):2179-2191

[23] Ryoo IG, Kwak MK. Regulatory crosstalk between the oxidative stress-related transcription factor Nfe2l2/Nrf2 and mitochondria. Toxicology and Applied Pharmacology. 2018;**359**:24-33

[24] Ungvari Z, Bailey-Downs L, Gautam T, Sosnowska D, Wang M, Monticone RE, et al. Age-associated vascular oxidative stress, Nrf2 dysfunction, and NF-{kappa}B activation in the nonhuman primate Macaca mulatta. The Journals of Gerontology. Series A, Biological Sciences and Medical Sciences. 2011;**66**(8):866-875

[25] Ahmed SM, Luo L, Namani A, Wang XJ, Tang X. Nrf2 signaling pathway: Pivotal roles in inflammation. Biochimica et Biophysica Acta (BBA) – Molecular Basis of Disease. 2017;**1863**(2):585-5971

[26] Ahn C-B, Je J-Y, Kim Y-S, Park S-J, Kim BI. Induction of Nrf2-mediated phase II detoxifying/antioxidant enzymes in vitro by chitosan-caffeic acid against hydrogen peroxide-induced hepatotoxicity through JNK/ERK pathway. Molecular and Cellular Biochemistry. 2016;**424**(1-2):79-862

[27] Itoh K, Wakabayashi N, Katoh Y, Ishii T, Igarashi K, Engel JD, et al. Keap1 represses nuclear activation of antioxidant responsive elements by Nrf2 through binding to the amino-terminal Neh2 domain. Genes & Development. 1999;**13**(1):76-863

[28] Kensler TW, Egner PA, Davidson NE, Roebuck BD, Pikul A, Groopman JD. Modulation of gene expression by cancer chemopreventive dithiolethiones and their metabolites. Journal of Biological Chemistry. 1995;**270**(27):16564-16571

[29] Kwak M-K, Itoh K, Yamamoto M, Kensler TW. Enhanced expression of the transcription factor Nrf2 by cancer chemopreventive agents: Role of antioxidant response element-like sequences in the Nrf2 promoter. Molecular and Cellular Biology. 2002;**22**(9):2883-2892

[30] Rushmore TH, Pickett CB. Glutathione S-transferases, structure, regulation, and therapeutic implications.

Journal of Biological Chemistry. 1990;**265**(30):17301-17304

[31] Lewis KN, Mele J, Hayes JD, Buffenstein R. Nrf2, a Guardian of Healthspan and Gatekeeper of species longevity. Integrative and Comparative Biology. 2010;**50**(5):829-8431

[32] Zhang DD. Mechanistic studies of the Nrf2-Keap1 signaling pathway. Drug Metabolism Reviews. 2017;**48**(3):34-412

[33] Wang XJ, Sun Z, Villeneuve NF, Zhang S, Zhao F, Li Y, et al. Nrf2 enhances resistance of cancer cells to chemotherapeutic drugs, the dark side of Nrf2. Carcinogenesis. 2008;**29**(6):1235-12434

[34] Zhang DD. The Nrf2-Keap1-ARE signaling pathway: The regulation and dual function of Nrf2 in cancer. Antioxidants & Redox Signaling. 2019;**29**(17):1628-1639

[35] Hayes JD, Dinkova-Kostova AT. The Nrf2 regulatory network provides an interface between redox and intermediary metabolism. Trends in Biochemical Sciences. 2014;**39**(4):199-218

[36] Xue P, Hou Y, Zhang Q, Woods CG, Yarborough K, Liu H, et al. Nrf2 affects the efficiency of mitochondrial fatty acid oxidation. Biochemical Journal. 2013;**457**(3):415-424

[37] Hu D, Linders A, Yamak A, Correia C, Kijlstra JD, Garakani A, et al. Metabolic maturation of human pluripotent stem cell-derived cardiomyocytes by inhibition of HIF1α and LDHA. Circulation Research. 2018;**123**(9):1066-1079

[38] Li Y, Zhang C, Li X, Ding J, Wu K, Fan D, et al. The dual role of Nrf2 in nonalcoholic fatty liver disease: Regulation of antioxidant defenses and hepatic lipid metabolism. European Journal of Pharmacology. 2015;**755**:89-96

[39] Zhao J, Liu L, Li X, Zhang L, Lv J, Guo X, et al. Neuroprotective effects of an Nrf2 agonist on high glucose-induced damage in HT22 cells. Biological Research. 2019;**52**(1):1-11

[40] Tsushima M, Liu J, Hirao W, Yamazaki H, Tomita H, Itoh K. Emerging evidence for crosstalk between Nrf2 and mitochondria in physiological homeostasis and in heart disease. Archives of Pharmacal Research. 2020;**43**(3):286-296

[41] Kang T-C. Nuclear factor-erythroid 2-related factor 2 (Nrf2) and mitochondrial dynamics/Mitophagy in neurological diseases. Antioxidants. 2020;**9**(7):617

[42] Corsello T, Komaravelli N, Casola A. Role of hydrogen sulfide in NRF2-and sirtuin-dependent maintenance of cellular redox balance. Antioxidants. 2018;**7**(10):129

[43] Novak I. Mitophagy: A complex mechanism of mitochondrial removal. Antioxidants & Redox Signaling. 2012;**17**(5):794-802

[44] Gureev AP, Sadovnikova IS, Starkova NN, Starkov AA, Popov VN. p62-Nrf2-p62 mitophagy regulatory loop as a target for preventive therapy of neurodegenerative diseases. Brain Sciences. 2020;**10**(11):847

[45] Esteras N, Abramov AY. Nrf2 as a regulator of mitochondrial function: Energy metabolism and beyond. Free Radical Biology and Medicine. 2022;**189**:136-153

[46] Digaleh H, Kiaei M, Khodagholi F. Nrf2 and Nrf1 signaling and ER stress

crosstalk: Implication for proteasomal degradation and autophagy. Cellular and Molecular Life Sciences. 2013;**70**:4681-4694

[47] Arefin S, Buchanan S, Hobson S, Steinmetz J, Alsalhi S, Shiels PG, et al. Nrf2 in early vascular ageing: Calcification, senescence and therapy. Clinica Chimica Acta. 2020;**505**:108-118

[48] Murakami A. Hormesis-mediated mechanisms underlying bioactivities of phytochemicals. Current Pharmacology Reports. 2020;**6**:325-334

[49] Lekshmi VS, Asha K, Sanicas M, Asi A, Arya UM, Kumar B. PI3K/Akt/Nrf2 mediated cellular signaling and virus-host interactions: Latest updates on the potential therapeutic management of SARS-CoV-2 infection. Frontiers in Molecular Biosciences. 2023;**10**:1158133

[50] Shou J, Deng Q, Jiang L, Gu L, Wei L. Inhibition of PI3K/Akt/NF-κB signaling pathway enhances chemosensitivity to cisplatin in tongue squamous cell carcinoma cells. Cancer Biology & Therapy. 2020;**21**(4):346-354

[51] Lal R, Dharavath RN, Chopra K. Nrf2 signaling pathway: A potential therapeutic target in combating oxidative stress and neurotoxicity in chemotherapy-induced cognitive impairment. Molecular Neurobiology. 2023:1-16

[52] Al Taleb Z, Petry A, Chi TF, Mennerich D, Görlach A, Dimova EY, et al. Differential transcriptional regulation of hypoxia-inducible factor-1α by arsenite under normoxia and hypoxia: Involvement of Nrf2. Journal of Molecular Medicine. 2016;**94**:1153-1166

[53] Yu L, Wang Y, Guo YH, Wang L, Yang Z, Zhai ZH, et al. HIF-1α alleviates high-glucose-induced renal tubular cell injury by promoting parkin/PINK1-mediated Mitophagy. Frontiers in Medicine. 2022;**8**:803874

[54] Zimta AA, Cenariu D, Irimie A, Magdo L, Nabavi SM, Atanasov AG, et al. The role of Nrf2 activity in cancer development and progression. Cancers. 2019;**11**(11):1755

[55] Calabrese V, Scapagnini G, Davinelli S, Koverech G, Koverech A, De Pasquale C, et al. Sex hormonal regulation and hormesis in aging and longevity: Role of vitagenes. Journal of Cell Communication and Signaling. 2014;**8**:369-384

[56] Yu C, Xiao JH. The Keap1-Nrf2 system: A mediator between oxidative stress and aging. Oxidative Medicine and Cellular Longevity. 2021;**2021**:1-16

[57] Pall ML, Levine S. Nrf2, a master regulator of detoxification and also antioxidant, anti-inflammatory and other cytoprotective mechanisms, is raised by health promoting factors. Sheng Li Xue Bao. 2015;**67**(1):1-18

[58] Chambel SS, Santos-Gonçalves A, Duarte TL. The dual role of Nrf2 in nonalcoholic fatty liver disease: Regulation of antioxidant defenses and hepatic lipid metabolism. BioMed Research International. 2015;**2015**:597134. DOI: 10.1155/2015/597134. Epub 2015 May 18. PMID: 26120584; PMCID: PMC4450261

[59] Li J, Xu C, Liu Q. Roles of NRF2 in DNA damage repair. Cellular Oncology. 2023;**46**(6):1577-1593

[60] Medoro A, Saso L, Scapagnini G, Davinelli S. NRF2 signaling pathway and telomere length in aging and age-related diseases. Molecular and Cellular Biochemistry. 2023:1-17. DOI: 10.1007/s11010-023-04878-x

[61] Matović V, Buha A, Đukić-Ćosić D, Bulat Z. Insight into the oxidative stress induced by lead and/or cadmium in blood, liver and kidneys. Food and Chemical Toxicology. 2015;**78**:130-140

[62] Cui T, Lai Y, Janicki JS, Wang X. Nuclear factor erythroid-2 related factor 2 (Nrf2)-mediated protein quality control in cardiomyocytes. Frontiers in Bioscience (Landmark edition). 2016;**21**:192

[63] Chapple SJ, Siow RC, Mann GE. Crosstalk between Nrf2 and the proteasome: Therapeutic potential of Nrf2 inducers in vascular disease and aging. The International Journal of Biochemistry & Cell Biology. 2012;**44**(8):1315-1320

[64] Zhang L, Ma Q, Wang Z, Wu Z. The relevance of Nrf2 pathway and autophagy in pancreatic cancer cells upon stimulation of reactive oxygen species. Pancreatology. 2016;**1**(16):S37. DOI: 10.1155/2016/3897250

[65] Jiang T, Sun Q, Chen S. Oxidative stress: A major pathogenesis and potential therapeutic target of antioxidative agents in Parkinson's disease and Alzheimer's disease. Progress in Neurobiology. 2016;**147**:1-19

[66] Lee J, Jang J, Park SM, Yang SR. An update on the role of Nrf2 in respiratory disease: Molecular mechanisms and therapeutic approaches. International Journal of Molecular Sciences. 2021;**22**(16):8406

[67] Bellanti F, Vendemiale G. The aging liver: Redox biology and liver regeneration. Antioxidants & Redox Signaling. 2021;**35**(10):832-847

[68] Mohamed EM, Samak MA. Therapeutic potentials of mesenchymal stem cells on the renal cortex of experimentally induced hypertensive albino rats: Relevant role of Nrf2. Tissue and Cell. 2017;**49**(2):358-367

[69] Abdelsalam HM, Samak MA, Alsemeh AE. Synergistic therapeutic effects of Vitis vinifera extract and Silymarin on experimentally induced cardiorenal injury: The pertinent role of Nrf2. Biomedicine & Pharmacotherapy. 2019;**110**:37-46

[70] Liu Q, Gao Y, Ci X. Role of Nrf2 and its activators in respiratory diseases. Oxidative Medicine and Cellular Longevity. 2019;**2019**:7090534-7090534

[71] Schiffers C, Reynaert NL, Wouters EF, van der Vliet A. Redox dysregulation in aging and COPD: Role of NOX enzymes and implications for antioxidant strategies. Antioxidants. 2021;**10**(11):1799

[72] Kourakis S, Timpani CA, de Haan JB, Gueven N, Fischer D, Rybalka E. Targeting Nrf2 for the treatment of Duchenne muscular dystrophy. Redox Biology. 2021;**38**:101803

[73] Frantz MC, Rozot R, Marrot L. NRF2 in dermo-cosmetic: From scientific knowledge to skin care products. BioFactors. 2023;**49**(1):32-61

[74] Hohmann MS, Zaninelli TH, Staurengo-Ferrari L, Manchope MF, Badaro-Garcia S, de Freitas A, et al. Nrf2 in immune responses during inflammation. In: Nrf2 and Its Modulation in Inflammation. New York City: Springer International Publishing; 2020. pp. 23-49

[75] Yang R, Zhang J, Li J, Qin R, Chen J, Wang R, et al. Inhibition of Nrf2 degradation alleviates age-related osteoporosis induced by 1, 25-Dihydroxyvitamin D deficiency. Free Radical Biology and Medicine. 2022;**178**:246-261

[76] Yan J, Li J, Zhang L, Sun Y, Jiang J, Huang Y, et al. Nrf2 protects against acute lung injury and inflammation by modulating TLR4 and Akt signaling. Free Radical Biology and Medicine. 2018;**121**:78-85

[77] Nenkov M, Ma Y, Gaßler N, Chen Y. Metabolic reprogramming of colorectal cancer cells and the microenvironment: Implication for therapy. International Journal of Molecular Sciences. 2021;**22**(12):6262

[78] Panieri E, Buha A, Telkoparan-Akillilar P, Cevik D, Kouretas D, Veskoukis A, et al. Potential applications of NRF2 modulators in cancer therapy. Antioxidants. 2020;**9**(3):193

[79] Wang R, Liang L, Matsumoto M, Iwata K, Umemura A, He F. Reactive oxygen species and NRF2 Signaling, friends or foes in Cancer? Biomolecules. 2023;**13**(2):353

[80] Qin S, He X, Lin H, Schulte BA, Zhao M, Tew KD, et al. Nrf2 inhibition sensitizes breast cancer stem cells to ionizing radiation via suppressing DNA repair. Free Radical Biology and Medicine. 2021;**169**:238-247

[81] Yamadori T, Ishii Y, Homma S, Morishima Y, Kurishima K, Itoh K, et al. Molecular mechanisms for the regulation of Nrf2-mediated cell proliferation in non-small-cell lung cancers. Oncogene. 2012;**31**(45):4768-4777

[82] Guo Z, Mo Z. Keap1-Nrf2 signaling pathway in angiogenesis and vascular diseases. Journal of Tissue Engineering and Regenerative Medicine. 2020;**14**(6):869-883

[83] Gao B, Doan A, Hybertson BM. The clinical potential of influencing Nrf2 signaling in degenerative and immunological disorders. Clinical Pharmacology: Advances and Applications. 2014;**6**:19-34. DOI: 10.2147/CPAA.S35078

[84] Telkoparan-Akillilar P, Panieri E, Cevik D, Suzen S, Saso L. Therapeutic targeting of the NRF2 signaling pathway in cancer. Molecules. 2021;**26**(5):1417

[85] Xue D, Zhou X, Qiu J. Emerging role of NRF2 in ROS-mediated tumor chemoresistance. Biomedicine & Pharmacotherapy. 2020;**131**:110676

[86] Raghunath A, Sundarraj K, Arfuso F, Sethi G, Perumal E. Dysregulation of Nrf2 in hepatocellular carcinoma: Role in cancer progression and chemoresistance. Cancers. 2018;**10**(12):481

[87] Choi BH, Kwak MK. Shadows of NRF2 in cancer: Resistance to chemotherapy. Current Opinion in Toxicology. 2016;**1**:20-28

Chapter 2

Scope of Theoretical Approaches Studying the Ligand-Nrf2 Interactions

Marvin A. Soriano-Ursúa, Ana L. Ocampo-Néstor, Melvin N. Rosalez, Eunice D. Farfán-García, Eduardo Osiris Madrigal-Santillán, Ángel Morales-González and José Antonio Morales-González

Abstract

The nuclear factor erythroid 2-related factor 2 (Nrf2) is a protein transcriptional factor considered as an emerging target for multiple biological processes. Its study could be useful to develop new treatments. Theoretical approaches have led to investigating the relevance of nuanced details in the structure of that protein. Also, after the X-ray crystals reporting Nrf2-KEAP1 interactions and identification of sites for potential interaction of other proteins and small compounds, the number of reports regarding these interactions is increasing. In this chapter, the advancements of theoretical approaches were collected and presented as useful to apply for developing therapeutic tools. First, information regarding the structure of Nrf2, as well as the observed interactions in crystal complexes is presented and commented. Thereafter, the identified sites of interest for letting interactions of natural compounds or small drugs which are proposed as potential drugs are presented. Finally, the advances and limitations of the theoretical assays, as well as the relationship with data reported experimental assays and the potential applications in medicine, are stated. Updated advancements by application of theoretical approach for understanding ligand interactions on Nrf2 are helpful to understand the biological activity of Nrf2 and the origin of some fails involved in etiology and pathophysiology of some human diseases.

Keywords: Nrf2, antioxidants, molecular modeling, proteins, Nrf2 activators

1. Introduction

However, if nuclear factors and their actions had been explored in the 1980s, the nuclear factor erythroid 2-like 2 (Nrf2) was cloned in 1994, and the studies detailing the molecular interactions that drive Nrf2 signaling were after its cloning [1]. Now, 30 years after that discovery, it is recognized as one of the major modulators of antioxidant response elements (ARE) and a mediator of the resolution of inflammation.

IntechOpen

Both hyperactivation and hypoactivation of Nrf2 contribute to the onset of chronic inflammatory processes, which are often linked to chronic diseases or cancer.

Whereas the detailed history of progress in the Nrf2 field is beyond the content of this chapter, some key facts are highlighted in a timeline (**Figure 1**). In brief, the domains of Nrf2 were described as it was cloned [2]; it was determined that Nrf2 is negatively regulated at the protein level by Kelch-like ECH-associated protein 1 (KEAP1) [3]. Also, the first Nrf2−/− mouse was born, and it was observed with disrupted antioxidant systems, such as a significant reduction in the levels of GST [4].

Then, the chemoprotective role of Nrf2 was suggested, and ways to activate the Nrf2 emerged as the target. Later, it was discovered that KEAP1-independent degradation of Nrf2 can occur because of glycogen synthase kinase β (GSK3-β) phosphorylation of key residues in Nrf2 [5, 6]. In 2006, the discovery that KEAP1 is mutated in non-small lung cell carcinomas was the first evidence that Nrf2 could contribute to cancer progression and chemoresistance [7]. Thus, it emerged approaches about the role of inhibition of high Nrf2 expression to treat certain cancers. But additional data supported a "dark side of Nrf2," as examples are the description of autophagic dysfunction as a result of p62-dependent sequestration of KEAP1 [8, 9] and the activation of Nrf2 promoting recurrence of dormant tumor cells.

But the role of Nrf2 transcriptional factor is not limited to its activity as a modulator of antioxidative mechanisms but also as a regulator of cell cycle and proliferation in cancer (including pathways not modulated by oxidative stress). It is an emerging target for multiple biological processes. It has been involved in processes that could be related to understanding the origin of maladies with high global burdens, such as neurodegeneration, inflammation, and cancer, and its study could be useful in developing new treatments.

Besides, it is clear that there is increasing interest in this study. Evidence of this can be observed in **Figure 2**, which shows the results of papers deposited in Pubmed in the last 30 years about Nrf2-studies. Both, academic research groups and pharma industry have increased efforts in understanding Nrf2 function and interactions.

Among the fields with the expansion of the knowledge of the Nrf2 details and functions, some specifications are involved in the origin and progress of some human maladies. But also, some research groups are focused on details of

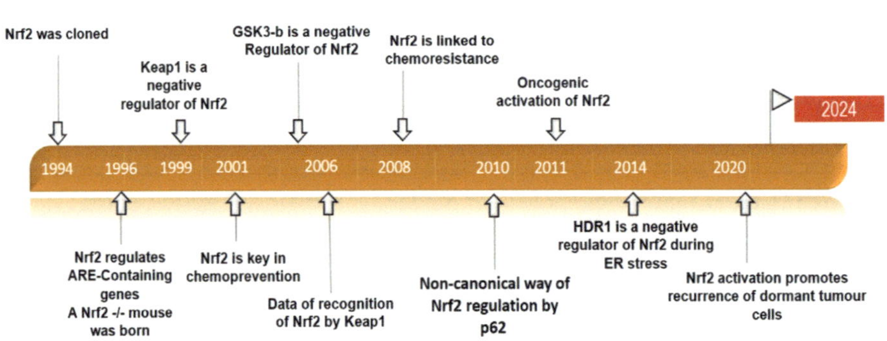

Figure 1.
Timeline of Nrf2 key reports about its function and interactions with proteins and ligands.

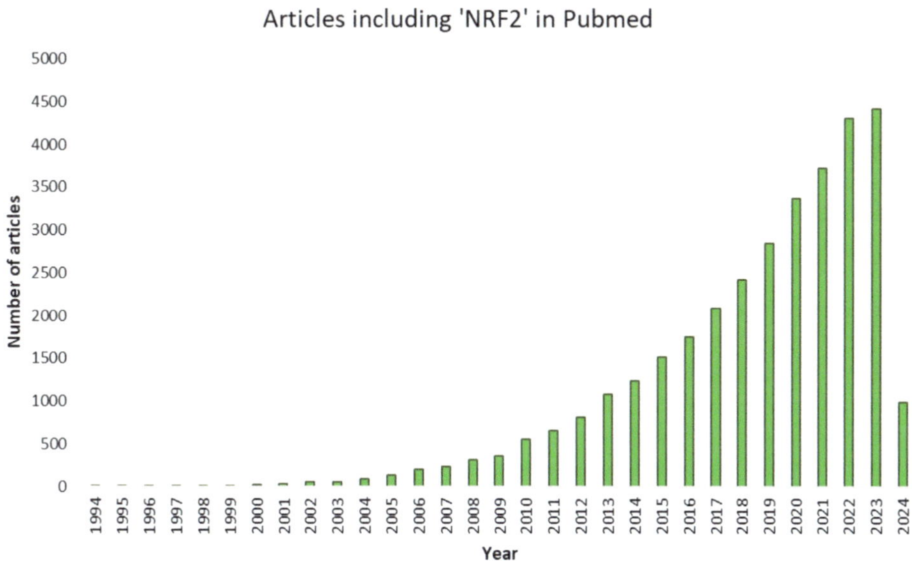

Figure 2.
Articles in Pubmed database including the term 'Nrf2' (by search on February 29, 2024).

interactions of Nrf2 with ligands or other proteins. After the X-ray crystals reporting Nrf2-KEAP1 interactions and identification of sites for potential interaction of other proteins and small compounds, the number of reports regarding these interactions is also increasing [1].

In this chapter, we have collected information regarding the structure of Nrf2 and the observed interactions in crystal complexes (of it, or those proteins reported as interacting with it). But also, we present the identified sites of interest for letting interactions of small drugs, which are proposed as potential drugs, and the results for quantitative structure-activity relationship assays and docking simulations, suggesting the properties of some compounds that could have those interactions on Nrf2.

Hence, the details of some residue movements which are suggested as key for ligand and protein interactions from molecular dynamics assays, are compiled. But also, the advancement of the theoretical assays, as well as the relationship with experimental assays and the potential applications in medicine commented.

2. More than anticancer drugs: Emerging applications of Nrf2 in biological processes and diseases

As mentioned above, the deepest study of Nrf2 has explained its action as an antioxidant regulator, mainly about the control of reactive oxygen species (ROS) actions and oxidative stress. As ROS have key roles in a wide range of physiological and pathophysiological processes, Nrf2 has been implicated in many human diseases with high global impact. Maybe the first attractive actions were in the regulation of cellular cycle and cancer [10]; however, involvement of the Keap1–Nrf2 pathway in tissue and organism homeostasis goes far beyond protection from cellular stress. Thus, emerging applications were from neurodegenerative diseases and neuropsychiatric disorders to cardiovascular disease and diabetes [11].

In fact, antioxidant defense systems have evolved as protective mechanisms against oxidative stress related to several diseases; being the Nrf2 a key regulator, involved in the detoxification and elimination of oxidative stress [12].

Accumulating evidence exists for oxidative stress and the Nrf2 pathway dysfunction during the origin and development of several neurological, metabolic, and inflammatory disorders, including Alzheimer's disease, Parkinson's disease, epilepsy, lateral amyotrophic sclerosis, diabetes, obesity, inflammatory bowel disease, and autoimmune diseases [13, 14].

Thus, the beneficial role of current molecular Nrf2 agonists/antagonists is evident, and their increasing use in ongoing clinical trials is expected. In fact, both activation and inactivation of Nrf2 are related to potential treatment, making Nrf2 a promising target for regulation of numerous diseases. In this sense, multiple natural or synthetic small compounds (the most cyclic compounds or peptides) have shown activity as agonists of Nrf2, acting directly on it, or by inhibition of the turning-off action of Keap1. Conversely, several other compounds have induced inhibition of Nrf2 by means of blocking dynamic key of Nrf2 or disrupting protein-protein interactions [15, 16]. However, further studies are required to explore the role of Nrf2 in the pathogenesis of diseases and suggest safe regulation before therapeutic use in clinical practice.

In spite of this, interest in clinical application is evident, as currently in clinical trials, 124 registers were found (on clinicaltrials.gov, searched on August 26, 2024) targeting neurodegenerative, vascular, inflammatory, and kidney diseases, most by modulation of Nrf2 by sulforaphane, broccoli sprouts or resveratrol administration, but also by specific Nrf inhibitors (such as pyrimethamine or telaglenastat) and activators (such as omaveloxolone or ebselen). Details can be found at: https://classic.clinicaltrials.gov/ct2/results?cond=&term=Nrf2.

In fact, today, Nrf2 is among the first targets studied by classic and systems medicine approaches to facilitate drug development by focusing on a set of diseases linked by a long protein complex (as that of Nrf2 and the protein's interaction with it). Those studies also have yielded the suggestion that some known drugs can be repurposed on diseases. Different approaches have been applied to find the alterations of Nrf2 expression and activity as a main mechanism in diseases. Thus, heterogeneous diseases such as autoimmune, respiratory, digestive, cardiovascular, metabolic, and neurodegenerative diseases, along with cancer, have been reported as potentially modifiable by targeting the Nrf2.

As an example, Cuadrado et al. reported an approach that matches and confirms in silico several applications for Nrf2-modulating drugs validated *in vivo* at different phases of clinical development. The chemical structures of those compounds suggested as ligands (the Nrf2-drugome), and their pharmacological profiles are as diverse as electrophilic dimethyl fumarate, synthetic triterpenoids like bardoxolone, sulforaphane, metformin, and statins as Nrf2 activators. The resulting selected drugs may rapidly expand the clinical options for human diseases [17].

Probably due to the high global burden of chronic-degenerative diseases, research on Nrf2 is biased to application in metabolic disorders, neurodegenerative maladies, and cancer. However, the first attractive clinical applications related to understanding of structural details of both, the ligand and target, were for bardoxolone and BG-12. The results of clinical trials for chronic kidney disease in type II diabetics were reported in 2011. After 52 weeks, the estimated glomerular filtration rate in the bardoxolone, 75 mg/day treatment group increased by about 32% when compared to entry values. That suggested Nrf2 activation represents a viable therapeutic approach for renal disease. On the other hand, patients with multiple sclerosis treated with BG-12 for 24 weeks showed fewer (32%) new gadolinium-enhancing lesions, with a significantly reduced probability of their evolution to T1-hypointense lesions than patients treated with placebo. The beneficial effects of these two "Nrf2 activators" triggered a new class of drugs to target many non-infectious diseases including cancer, cardiovascular diseases, inflammatory and autoimmune disorders, as well as neurodegenerative diseases [18].

The influence of Nrf2 on ROS and the way Nrf2 regulates antioxidant defense have attracted attention for application in neurodegenerative and metabolic disorders. For elucidating the mechanism(s) of action, multiple studies have been carried out, yielding information about different proteins and domains (of Nrf2) for specific interactions (**Table 1**).

At this point, it is adequate to keep in mind that the ability of the neuron to adapt against oxidative stress and maintain a clearance mechanism for misfolded aggregating proteins determines its survival, and it has been related to neurodegenerative processes. In this sense, increasing evidence supports the Nrf2-antioxidant response element pathway for mitigating several pathologic mechanisms associated with Alzheimer's disease, Parkinson's disease, amyotrophic lateral sclerosis, Huntington's disease, and multiple sclerosis [19].

Also, other proteins have been mentioned as modulating the action of Nrf2 by direct interactions (**Table 2**), but some others have attracted attention as modulators of Nrf2 interaction with key proteins, such as Keap1. Thus, p62, DPP3 WTX, and PALB2/FANCN act as Keap1 inhibitors disrupting direct Nrf2-Keap1 interaction.

Domain	Protein interacting	Function
Neh1	sMaf	Dimerization let interaction with DNA
Neh1	ARE	DNA-binding
Neh2	Keap1	Negative regulation
Neh3	CHD6	Transactivation
Neh4	CBP	Transactivation
Neh5	CBP	Transactivation
Neh6	βTrCP	Negative regulation
Neh7	RXRα	Suppressed transactivation

Table 1.
Some functional domains of Nrf2 and their respective binding proteins.

Protein	Nrf2-domain	Regulation effect
p21	Neh2	Nrf2 stabilization
BRCA1	Neh2	Nrf2 stabilization
βTrCP	Neh6	Inhibits Nrf2 degradation
RXRα	Neh7	Inhibits transactivation

Table 2.
Examples of proteins acting as Nrf2 regulators by direct interaction.

3. Nrf2-ligands interactions: From modeling, crystals, and docking studies

Several approaches have been used to study the structure and interaction of Nrf2, proteins and ligands interaction with this protein. Firstly, due to the inexistence of Nrf2-structures, using homology models supported by the identity of Nrf2 peptide sequence with other proteins. But interesting reports started in this century, before the crystallization of some fragments of Nrf2, or most residues conforming the KEAP1 or other Nrf2-related proteins coinciding with reports of involvement of Nrf2 in many cellular functions.

As an example, the initial exploration of Nrf2 structure and its specific interaction is the first crystallization of the Kelch domain of Keap1 and its clear linking to Nrf2 function in 2005 (available in PDB with ID: 1ZGK), thereafter, multiple structural analyses were done [20]. Those were increased after the identification of mutations disrupting interactions of ligands [21] and after the observation of complexes of Keap1 and fragments of Nrf2 due to the sites for protein-protein interactions and potential sites to disrupt those interactions were proposed or identified [22].

Williamson et al. [23] used the quantum model to perform *in silico* screening on over 18 million commercially available chemicals to identify the structures predicted to activate the Nrf2-antioxidant response element pathway based on the quantum model. To identify novel small molecules that will disturb Nrf2-Keap1 binding and promote activation of the Nrf2-antioxidant response element pathway, they generated a quantum model based on the structures of known Nrf2-antioxidant response element activators. The resulting improved ability to identify potent activators of Nrf2 through the combination of *in silico* and *in vitro* screening enhanced the speed and cost associated with screening Nrf2-antioxidant response element, then, triggering innovative and wide screening compounds for drug development [23].

Other relevant data, as it is key for advances in the Nrf2 function, were from crystallographic assays yielding an X-ray crystal structure of KEAP1 co-crystallized with (1S,2R)-2-[(1S)-1-[(1,3-dioxo-2,3-dihydro-1H-isoindol-2-yl)methyl]-1,2,3,4-tetrahydroisoquinolin-2-carbonyl] cyclohexane-1-carboxylic acid (compound (S,R,S)-1a). That X-ray crystal structure provides breakthrough experimental evidence for the true binding mode of a compound (S,R,S)-1a. But also, increased the interest in exploring more Nrf2-related proteins, as the ligand orientation was found to differ from that of the initial docking model, which was available at the initial phase of that project. Crystallographic elucidation of this binding mode helped to focus and drive the drug design process more effectively and efficiently [24].

Currently, more than 150 structures related to Nrf2 are deposited in the Protein Data Bank. Interactions between Nrf2 and KEAP1 remain the most reported cases. With those structures as templates, *in silico* drug development techniques have come a long way throughout the years. Another area that has seen drastic increases in development is high-throughput protein purification through crystallography and nuclear magnetic resonance spectroscopy, which have contributed to a better understanding of the interactions of protein-ligand complexes. *In silico* techniques help to better understand and study drug discovery and in comparison with the traditional high-throughput screening, these techniques are a more direct drug discovery approach with the added advantages of being very low cost and having an effective screening. Different programs use different algorithms to carry out molecular docking studies, allowing the analysis of a small molecule and a protein at the atomic level, that is, helping elucidate fundamental biochemical processes [25–27].

There are different types of interactions sharing Nrf2 as a part of complexes; as a classification, it can be said that there are protein-protein interactions (PPIs) and ligand-protein interactions [28, 29]. In the first case relevant are those interactions between the Nrf2 and KEAP1. The Nrf2 belongs to the Cap'n´Collar (CNC) subfamily of basic leucine zipper (bZIP) transcription factors and possesses seven conserved Nrf2—ECH homology (Neh) domains. However, many other proteins are linked to key interactions (see summary in **Table 1**). The Neh1 domain contains a bZIP structure for dimerization with small musculoaponeurotic fibrosarcoma proteins, Maf, K, G, F which help to recognize antioxidant response elements, ARE for activation of gene transcription and to bind to DNA. The Neh2 domain contains ETGE and DLG motifs that specifically bind to the Kelch domain of the KEAP1 protein for mediation of ubiquitination and degradation and is located at the N-terminal region. For the Neh3–5 domains, they work as transcriptional activation domains for binding of various components of the transcriptional machinery. Neh6 contains two degrons DSGIS and DSAPGS which bind to E3 ubiquitin ligase beta-transducin repeat-containing protein, βTrCP, which mediates Nrf2 degradation in oxidatively stressed cells. Lastly, the neh7 mediates the interaction with retinoic X receptor alpha, RXRα, which represses Nrf2 activity. Thus, is notable that it can be modulated at multiple levels [30].

Thanks to the analysis of crystallized structures, it is now well-known that KEAP1 belongs to the BTB (Broad Complex, Tramtrack, and Bric-a-Brac,-Kelch family), which contains five domains, its amino-terminal region, a broad complex, tramtrack, bric-a-brac domain, an intervening region, IVR, six Kelch domains, and the C-terminal region. The Kelch domains contain six Kelch repeats that fold into six bladed-β-propeller structure, with each blade having a four-stranded antiparallel β sheet (A-D) in which shorter βA strands form the central core with the final βA strand from the C-terminal closing the propeller by completing blade 1. Although diverse in sequence, they do maintain conserved positions that maintain their fold. These regions include a double-glycine repeat, DGR that terminates the βB strand as well as individual tyrosine and tryptophan residues that mediate hydrophobic packing between blades. The substrate binding surface lies on one face of the Kelch Domain, where an area is created between long loops connecting β strands D and A, DA loop, and β-strands B and C, BC loop. The bound ETGE peptide of Nrf2 adopts a β-turn conformation inserted into this pocket and, with specific electrostatic interactions, is made by glutamate residues in the ETGE motif. Glu79 forms residues with Arg415, Arg483, and Ser508, whereas Glu82 forms hydrogen bonds with Ser363, Asn 383, and Arg380. The DLG motif of Neh2 of Nrf2 is an independent binding site but is 100-fold weaker affinity for KEAP1 but did show a β-turn conformation similar to ETGE with an N-terminal helix, Leu19-Arg25, and two short 3 10 helices, Ile28 to Leu30 and Arg34 to Phe37, with residues Asp29-Leu30-Gly31 being located similar to ETGE's Glu79-Thr80-Gly81. The BTB domain homodimerizes and contributes to interacting with IVR and the CRL3 class of Cullin-RING-Ligase E3 ligases that provide recruitment for ubiquitination with special interest in the cysteine residues C151, C226, C273, and C278 that sense ROS and electrophiles and modulate the interaction of Keap1/Nrf2 interactions [31].

On the other hand, in the case of ligand-protein complexes, multiple studies have been carried out. In this sense, phytochemicals have had a great impact in medicine, showing a mitigation or prevention of chronic disease in humans and various studies have shown that they counteract oxidative stress. Thus, phenolic compounds act as antioxidants which scavenge free radicals, thereby delaying or interrupting the propagation step of lipid oxidation. Vitamin E acts as a peroxyl radical-scavenging antioxidant and as an inhibitor of lipid peroxidation *in vitro* and in vivo, while carotenoids act as chemical quenchers. Most of these compounds even exhibit their effect through the Nrf2 pathway, and their interactions on Nrf2 and related proteins are often suggested or probed.

As an example, in a recent study, 178 phytochemicals were used, with the previous search using Pubmed, Web of Science, and Google Scholar. Their 3D structure was downloaded as SDF from the PubChem Chemical library, while the KEAP1 structure code 417b was used. They were prepared using the SYBYL X 2.0 software. The Surflex-Dock module was used with protomol generation using a ligand-based approach with 3D-quantitative structure-activity relationship (QSAR) studies, cell culture, and cell viability assay. For the analysis, the C-Score was used, which shows hydrogen bonding, metal and ligand interaction, lipophilic contact, and rotational entropy while the Total_Score shows the binding affinity between the ligand protein, with a Total_Score > 6 having ligand candidate for further study and C-Score > 4 the candidate being more promising. Of all compounds, 24 showed a binding affinity with KEAP1 > 4 and interacting with amino acids including Tyr334, Ser363, Lys365, Asn383, Ser383, Pro384, Asp385, Asn414, Ile416, Val463, Ser508, Ser555, Leu557, Tyr572, and Ser602. These amino acids have been reported at other times as relevant in the Nrf2 activities (**Figure 3**).

With the prediction of the binding affinity values, both the presence of glycosides and number of oxygen molecules are suggested as key in the ligands for interactions on Nrf2. The compounds were divided into three groups depending on their affinity, and 11 were chosen to evaluate their activation effect on Nrf2 including cyanidin 3-sambubioside, C3S, luteolin-5-O glucoside, L5G, rutin, RUT, apigenin, API, kaempferol, KAEM, α-tocopherol, αT, coptisine, COP, magnoflorine, MAG, piperine, PIP, ellagic acid, EA and echinacoside, ECH. With the cell viability assay, Western blot, and gray density, the studies showed these compounds have no significant influence on VRF2 expression in the cytosol, while they showed expression of Nrf2 in the nucleus, showing that even under oxidative stress, the compounds did show potential binding affinity to the Nrf2 binding site in KEAP1 and thus activating Nrf2 [32].

Other similar molecules and methods have been employed. Thus, another study was carried out with citrus-derived flavonoids that have shown bioactive functions such as antioxidants against inflammation, allergy, cancer, and lowering blood

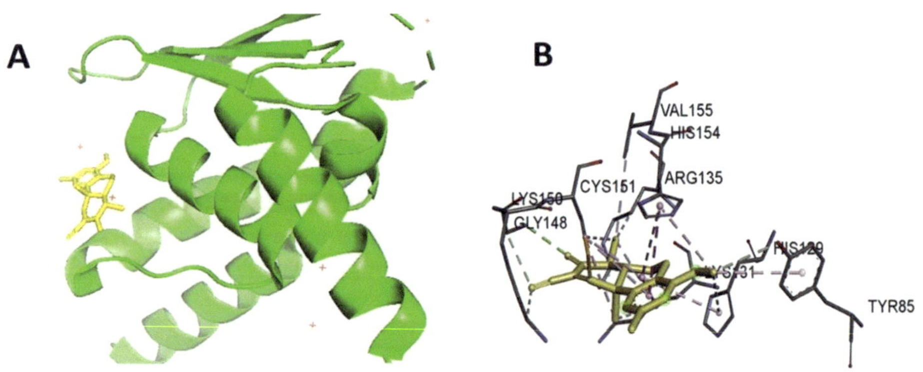

Figure 3.
Depiction of Human BTB Domain of KEAP1 (A) Small potential inhibitor interacting with BTB Domain. (B) Main Residues interacting with small inhibitor including Cys151.

lipids and cholesterol. The compounds naringenin, hesperetin, narirutin, naringin, hesperidin, neohesperidin, nobiletin, neohesperidin dyxdochalcone and tert-butylhydroquinone were used and Density Functional Theory was chosen for theoretical computations, with B3LYP hybrid functional with standard 6-311G basis set with a ++ diffuse function for geometry optimization using Gaussian 09 and Gaussview 06 to obtain their highest occupied molecular orbital (HOMO), and lowest occupied molecular orbital (LUMO) providing details on the binding behavior with the receptor. In the next step, the crystallized structure of the protein, the PDB ID 1X2R, was selected and using Autodock Vina as well as a specific grid box spacing of 0.375 Å was generated, which encircled all critical residues of KEAP1, and the lowest binding energy (kcal/mol); then, interactions was calculated based on the Lamarckian genetic algorithm and the compounds were visualized on Pymol and Discovery Studio 2018. For drug-likeness and toxicity, the canonical simplified molecular input line entry, SMILE, was fed to the SwissADME server. The topology of the flavonoids showed that most electrophilic attacks occurred in the C=O. With the data obtained on the docking software, the compounds were divided as a weak agonist, naringin, neohesperidin, dihydrochalcone, nobiletin; and strong agonist, narirutin, naringin, hesperetin, and hesperidin. The results showed that the tested compounds bind to the binding pocket of KEAP1, with nargiring interacting with hydrogen bonds with Val512 and Val465, while hesperetin also formed hydroxyl groups with Gly367, Val606, and Arg415. The weak agonists bound to Arg380, Ser431, Arg483, Asn414, Arg415, Leu365, and Tyr572 with multiple hydrogen bonds as well as Pi-Alkyl and Pi-Sigma helping stability and closeness between ligand and KEAP1 [33].

A study with synthetic compounds as potential ligands, based on previous studies supporting bis-acidic naphthalene derivatives as the core was done. Then, a series of xylene-based bis-sulfonamides as direct inhibitors of the KEAP1-Nrf2 PPI with improved metabolic stability and using Autodock 4.0 software with the protein ID: 4XMB. The study found that the positional modification of two substitutions on the xylylene core with a 1,2-disubstitution, having a compound named 9A in the original manuscript, was optimal for substitution and the introduction of the S-methyl group on the alkyl side chain (in the compound 21a), showed a promising increase in activity in fluorescence Polarization assays. With regard to the docking results, both disubstituted compounds did interact with Arg415, Asn414, Arg483, and GLn530, with mainly hydrogen bonds formed. These compounds also showed higher metabolic stability in human liver microsomes; thus showing that xylylene compounds could be used as leads for further optimization in search for potent direct inhibitors of KEAP1-Nrf2 PPI [34].

Another advance in the search for disrupters of KEAP1-Nrf2 PPI was made. That study found that for the interaction of non-covalent KEAP1 inhibitors, a macrocycle, its ring size, the stereochemistry of the amino acid constituents, and dimethyl amide are key. From a base compound, they optimized other compounds that showed kI <100 nm. Those compounds that were located deeper in the binding site and formed a salt bridge with Arg483 showed a 100-fold increase in potency. They also form polar interactions with Arg483, highlighting the essential amino acid for high affinity. The phenylene ring of the derivatives showed that it was wedged in between Ala556 and Arg415, forming a pi-cation interaction, while the C-terminal dimethylamine stacked against Tyr572 while the para-benzoic acid moiety of the nM inhibitors formed a bidentate salt bridge with Arg483 and the carbonyl group of the lactam moiety interacts with Ser602 of KEAP1. They also showed that a limited reorientation of Arg415 and Arg483 can significantly affect the affinity of KEAP1 for a ligand (**Figure 4**) [35].

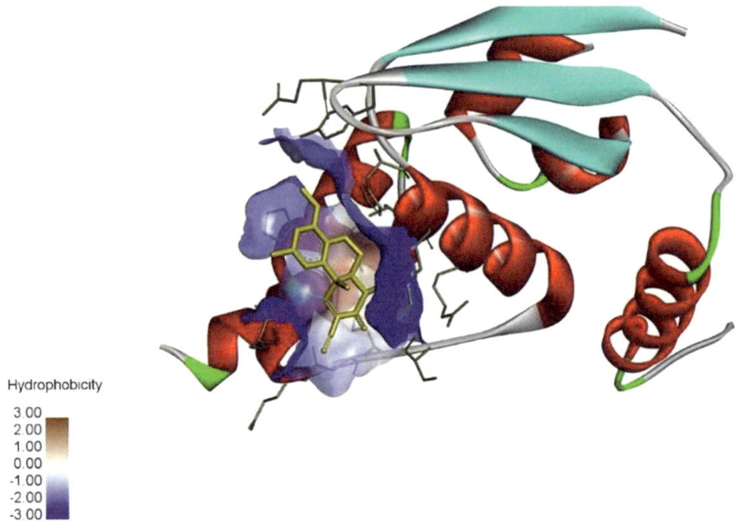

Figure 4.
Grade of Hydrophobicity of BTB Pocket and small inhibitor.

Thus, focusing on the protein-protein interaction is key in the study of Nrf2. Cumulative information lets us take plenty of assays of ligands suggested as modifiers of that interaction. The use of a PPI-oriented library comprising 12,593 small to medium-sized newly synthesized molecules was run through a ligand-based virtual screening. It was carried out to discover novel PPI inhibitory compounds for the KEAP1/Nrf2 interface with five hit compounds with Tanimoto similarity <0.26 and aqueous solubility, AlogP<5. The ligands were aligned 3D with the X-ray crystal structure of KEAP1/Nrf2 (PDB ID: 2flu) with a fit into the binding pocket of KEAP1, but also with Glu79, Glu82, Ser363, Arg380, Asn382, Arg415, Arg483 and Ser508 of the Nrf2. It was shown that for four of the hit compounds, a contributing aromatic ring was common for their substructure and interaction with Tyr572 of KEAP1. Also, some studies have focused on exploring the natural compounds with well-supported action explained by intervention in this KEAP1/Nrf2 interaction [29, 36, 37]. Other studies focus on protein-protein surface analysis by applying machine learning [38].

Although most aforementioned compounds tend to be small molecule inhibitors of KEAP1, there are also peptide inhibitors based on ETGE motif-containing peptides. Accordingly, 12 different peptides have been reported and shown through biological evaluation assay, including fluorescence polarization (FP) assay and competitive surface plasmon resonance (SPR) assay with different ki (considered as an affinity value) ranging from 389 nM to 6.1 nM. FP assay revealed the importance of the phosphate groups making polar interactions with Arg415, Arg483, and Ser508, increasing the contribution of the KEAP1 pocket [29].

Finally, it should be noted that some well-known drugs have been repurposed by using theoretical assays. Among these is the case of chlorhexidine, which could benefit patients with Parkinson's disease [39], and other complex molecules, such as some polymers; they have shown the ability to disrupt Nrf2 function and putative application in maladies, presumably by action on Nrf2 structure and its function [40].

4. QSAR studies of ligands for Nrf2

Due to the wide exploration of active compounds in Nrf2, the analysis of the structure of active compounds is a necessity. It has been approached through quantitative structure-activity relationship (QSAR) studies. Those let predictions of the activity of multiple compounds, even before they have been synthesized. Once QSAR models are generated, in addition to predicting activity, they can be used to better understand the structural requirements of compounds for them to exert their activity [41].

To obtain a good QSAR model it is essential to have a large training set and compounds that can be used to validate the model. The idea is to have a series of compounds in which the base structure is constant, and the structural changes are present in specific positions in order to relate the changes in activity with the changes in structure. An ideal QSAR model consists of at least one hundred compounds in the training set and approximately ten percent in the validation set. However, there is not always that number of compounds that, in addition to presenting structural similarity, have been experimentally evaluated on some biological activity of interest. The Nrf2-QSAR studies are no exception; they have been limited by the number of compounds, with several QSAR studies including training sets of just a dozen of compounds [41].

Furthermore, in various publications in which the activity mediated by Nrf2 is explored and in which QSAR studies are reported, a direct activity on this transcriptional factor is not always explored. But, activities related to it are determined, such as the antioxidant activity of the compounds of interest, which is validated through *in vitro* studies [42–45].

Despite the limited number of compounds in the training set, results have been obtained that guide the structural composition of the compounds that must present to exert better activity on Nrf2. Studies where theoretical QSAR results differ from *in vitro* results are typically studies with a limited range of screening compounds [44]. This is why, in the case of QSAR studies of Nrf2, as well as in other receptors, SAR studies become relevant, which allows obtaining experimental results of new compounds (often with 2 to 5 aromatic rings and oxygens atoms in their structure. Some details of compounds identified by QSAR studies as potentially applicable to specific diseases are in **Table 3**). These results can subsequently be used to expand the training sets of QSAR studies that have already been carried out and improve them or carry out new studies based on them [52].

On the other hand, there are also QSAR models to predict oxidative stress using Klf2(K Krüppel-like factor 2, which substantially enhances the antioxidant activity of Nrf2). They have been generated with a wide variety of chemical (> 4000) compounds but with unsatisfactory coefficients of determination values [51].

Also, it should be mentioned that a combination of computational methods, such as the use of 3D QSAR models, the search for chemical descriptors related to different targets, as well as extensive training sets, have resulted in the obtaining of compounds with a multitarget effect, with application in Alzheimer's disease, targeting the transcription factor Nrf2 [46].

Albeit if poorly explored QSAR studies for identifying peptides bioactive on Nrf2 are emerging. In fact, in the food industry, peptides of great importance have been obtained that can activate the keap1-Nrf2-ARE pathway and protect cells from oxidative stress accordingly with *in vitro* results. These peptides have been identified and selected by using computational tools to predict antioxidant activity and select the best candidates. Some of the chosen peptides have attracted attention as they can be obtained from Feng-flavor Baijiu [38] and from the *in vitro* digestion of pork sausage [48].

Medical application	QSAR characteristics
Alzheimer's disease	The compounds austrobailignan 6, anolignan c, 7-epi-virolin, 6-[(2R,3R,4R,5R)-3,4-dimethyl-5-(3,4,5-trimethoxyphenyl)oxolan-2-yl]-4-methoxy-1,3-benzodioxole, ococymosin, and mappiodoinin B have probabilities to confer neuroprotection and antioxidant activity via multitarget, including the target Nrf2 [46].
Parkinson's disease	Vinylsulfone disubstituted with benzene and pyrimidine derivatives were evaluated. It was found that in them, the H atoms at the ortho and meta positions of the pyrimidine have a positive effect on molecular activity. Furthermore, QSAR models demonstrated that the activity of the compounds can be improved by increasing the volume of the substituents on their benzene ring [47].
Treatment of cerebral ischemia-reperfusion injury (CIRI)	Novel monoenone monocarbonyl curcumin analogs with a potent protective effect by effectively eliminating the generation of ROS in response to oxidative stress injury through activation of the Nrf2/HO-1 antioxidant signaling pathway [45].
Effects of phthalate esters (PAEs) on cell viability and Nrf2 of Human hepatocyte HepG2	The introduction of large and hydrophobic groups may significantly affect the toxic effects of PAEs on cells [43].
Antioxidant effect	Tyr or Trp residues at the N terminus of dipeptides, tripeptides containing more Cys and Trp amino acid residues, and the presence of Tyr, Trp, Cys, and Met residues increase antioxidant activity by activating the keap1-Nrf2-ARE pathway [42, 48, 49]. The 3D-QSAR models suggest that a small and hydrophilic group in the N-terminal of the tripeptide or a bulky and hydrophobic group in the C-terminal were beneficial for bioactivity [50].
Oxidative stress induced by organic chemicals	The parameters nBM (number of multiple bonds) and H% (percentage of H atom) played dominant roles in the prediction of the activity (inactive or active) of chemicals [51].
SAR of peptides for the treatment of cancer	Inhibitors of the Nrf2/MAFG/DNA ternary complex. The introduction of the N-terminal loops Nrf2 and MAFG makes the peptides more potent due to a possible key role in the pre-organizing of the helical DNA-binding region of Nrf2 and MAFG [52].

Table 3.
Compendium of some QSAR studies of ligands of Nrf2.

Likewise, once the characteristics that favor antioxidant activity were established, peptidomimetic compounds were designed. Agreeing results, dipeptides containing Tyr, Trp, Cys, and Met residues play an important role in antioxidant activity. The Tyr or Trp residue at the N terminus of the dipeptides showed stronger antioxidant activities than when it is at the C terminus of the dipeptides. Furthermore, tripeptides containing more Cys and Trp amino acid residues exhibit higher antioxidant activities, and the steric properties of the amino acid residues at the N or C terminus as well as the electronic and hydrogen bonding properties of the amino acids are important for the antioxidant effect [49].

5. Molecular dynamics studies of Nrf2

Movements and folding of Nrf2 are interesting as they could be key for its activity. The movements in segments of the protein, as well as those movements after contact with other protein(s) or ligand(s), must be explored. In this context, molecular dynamics (MD) simulations of Nrf2 have yielded information.

Some examples of applications are mentioned in this section. Thus, inhibition of the Keap1/Nrf2 signaling pathway promotes the onset of type I diabetes (T1D) in experimental models, suggesting that finding drugs that can activate Keap1/Nrf2 signaling may be a promising therapeutic strategy for the origin and treatment of T1D. Formononetin (FMNT) could significantly increase the expression of Nrf2 and improve antioxidant response elements (AREs). Although FMNT could improve T1D, the mediating role of Keap1/Nrf2 signaling by FMNT in improving T1D is still unclear; experimental assays support modulation of folding in the target protein, while the interaction pattern between FMNT and Keap1 was assessed by MD simulation [53].

Through these types of computational studies, molecules with identified anti-inflammatory properties have been suggested as Nrf2 activators with potential application in drug-induced liver injury (DILI). That is interesting since, under oxidative stress, Nrf2 translocates to the interior of the nucleus after dissociating from KEAP1 and initiates the transcription of cytoprotective genes [54]. In 2023, Mili and co-workers presented a study in which they analyzed a database of more than 24,000 compounds with various computational techniques. Among the techniques to study Nrf2 interactions with ligands, KEAP1 or additional proteins involved in transcription, extra-precision molecular docking (molecular XP-docking) and MD were used [55].

The authors reported three top compounds (23,013, 12,895, and 9866 in order of predicted affinity) with anti-inflammatory activity, which act as Nrf2 activators to which they performed MD and showed predicted higher stability than the co-crystallized ligand (1VX, considered as reference-ligand on Nrf2 and found in the complex with PDB ID: 4L7D, details of its interaction are in **Table 4**). The analysis of predicted form of interaction for the selected compounds let them suggest that those compounds could disrupt the Nrf2 binding to KEAP1 in the cytoplasm and degradation by the Cul3-E3 ubiquitin ligase [55].

Some other compounds are proposed as possible drugs that act on Nrf2, and their action has been validated with *in vitro* experiments. However, their possible mechanisms have been explored in silico, and the results suggest action beyond Nrf2 as a unique receptor; that is, those compounds could also act in other nearby proteins (such as KEAP1, NF-kβ, or OH-1) [53, 56, 57]. Moreover, other compounds could regulate Nrf2-related processes, such as fructosamine-3-kinase (FN3K), which participates in the deglycation of Nrf2, which is important in the regulation of oxidative stress in cancer cells. Gefitinib, sorafenib, neratinib, tamoxifen citrate, cyclosporin A, and oxaliplatin modulated FN3K activity in breast cancer cell lines, with simultaneous alteration in antioxidant signaling mediated by Nrf2 [58]. Also, through MD studies,

Nrf2 is bound to these residues:	Compound 1VX, which acts as an activator of Nrf2 by binding with:	The best three compounds (23,013, 12,895, and 9866) show that they interacted with
SER363, ARG380, ASN382, ARG415, ARG 483, in ubiquitin ligase Cul3 [54].	SER363, ARG380, ASN382, ASN414, ARG415, and TYR572 residue of KEAP1 protein [24].	TYR 334, SER363, ASN 382, ASN414, ARG415, TYR572, and SER602 residue of the active pocket of KEAP1 protein [54]

Table 4.
Summary of key residues in the interaction to modulate DILI.

it was observed that the binding of Nrf2 with SIRT6 (one of the few proteins that can regulate both lifespan and aging) is mediated by the Neh1 and Neh3 domains, since the elimination of these domains decreased binding stability and free energy; some compounds and peptides could modulate Nrf2 action by binding in these protein segments [59].

The interaction of some natural products has also been evaluated by using MD simulations on Nrf2 and related proteins. Thus, the target proteins involved in diabetic nephropathy, such as the diabetic nephroprotective activity of Maquian essential oil (MQEO). α-Terpineol showed high binding affinity and formed a very stable complex with Nrf2, interacting with key residues such as Val 418 and 606, Gly 367, 511, 558, and 605, Ala 366 and 510, Leu 365 and 557, and Ile 416 and 559. In addition, it showed promising pharmacokinetic properties [60]. Similarly, various properties have been attributed to silymarin extract. Through MD simulations, it was observed that two flavonolignans (silybin A and 2,3-dehydrosilybin) in that

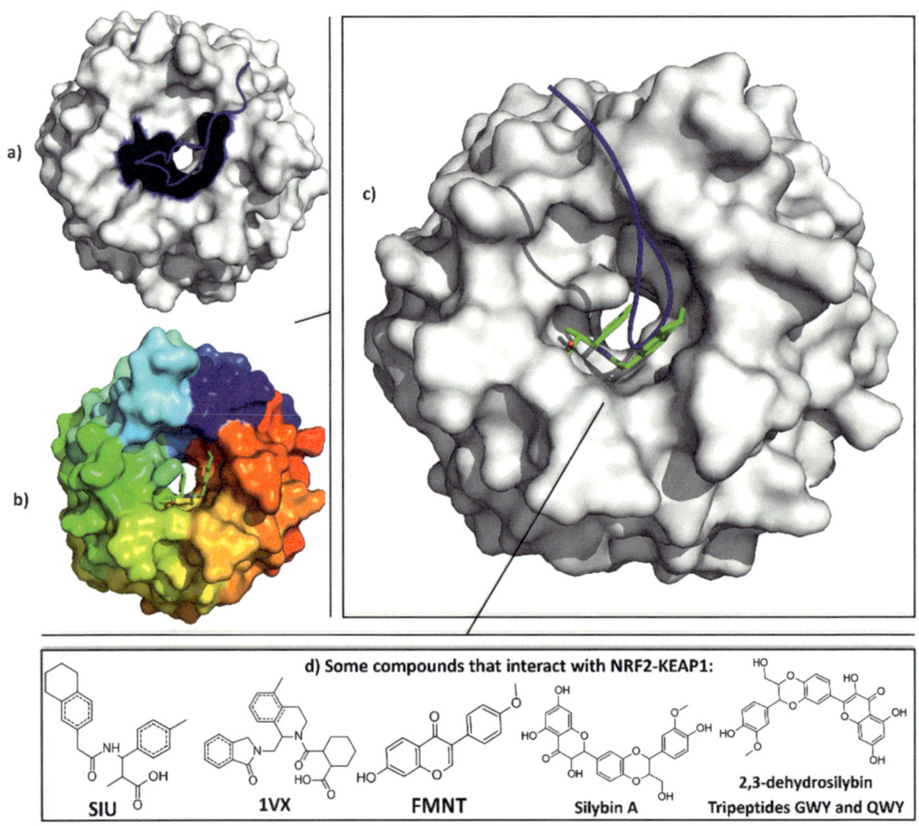

Figure 5.
Nrf2-Keap1 interactions and ligands disrupt them. a) Complex of the Neh2 domain (blue ribbon) of Nrf2 with Kelch (surface) of Keap1 (PDB:2FLU). b) The compound that is inserted into the Kelch domain is a ligand derived from propanoic acid (SIU), which represents the ligands that compete for the Nrf2-keap1 binding site (PDB: 8IVR). c) Merging the SIU and Neh2 interaction sites in Keap1 shows a common active site. d) Some compounds that interact with Nrf2-Keap1 as they have been mentioned in the present chapter: (2R,3S)-2-methyl-3-(4-methylphenyl)-3-[2-(5,6,7,8-tetrahydronaphthalen-2-yl)ethanoylamino]propanoic acid (SIU), (1S,2R)-2-{[(1S)-5-methyl-1-[(1-oxo-1,3-dihydro-2H-isoindol-2-yl)methyl]-3,4-dihydroisoquinolin-2(1H)-yl]carbonyl} ciclo-hexanecarboxylic acid (IVX), 7-hydroxy-3-(4-ethoxyphenyl)chromen-4-one (Formononetin, FMNT), Silybin A and 2,3-dehydrosilybin.

extract bind to Keap1, having interactions with the same pockets observed for the co-crystallized KEAP1-CPD16 complex, but with an increased number of interactions and higher affinity values, positioning these compounds as potential activators of Nrf2 [61].

On the other hand, MD has been applied to explore interactions of peptides modulating the binding of Keap1 with Nrf2. It has allowed us to identify antioxidant peptides, such as the GWY and QWY tripeptides, that could compete for the binding site with Nrf2 in agreement with observed movements and modification of protein motility and folding in the MD simulations. Hence, those peptides are attractive as activators of antioxidant systems (**Figure 5**) [50].

Finally, it should be mentioned that aberrant activation of Nrf2 in some cancer cells may confer cytoprotection of these cells. The key to this antioxidant pathway is through interactions with other proteins, such as Keap1, CREB-binding protein (CBP), and retinoid X receptor alpha (RXRα), which makes it an ideal target for the design of anti-cancer drugs. The interaction between Nrf2 and CBP is mediated by the Neh4 and Neh5 domains of Nrf2, and MD studies have shown that these domains are partially disordered [62]. Nrf2 is also known to display transiently structured elements in its Neh1, Neh2, and Neh7 domains. However, there is no complete structural information on these complexes, which limits the rational development of drugs considering these protein domains as targets [62, 63].

6. Conclusions and perspectives

Nrf2, its structure, the interactions with other proteins, and the putative recognition of ligands that modify its activity are attractive topics for developing new drugs to treat some diseases linked to Nrf2 dysfunction. Advances in the theoretical study of Nrf2 and its interactions to modulate biological processes are emerging. However, in just a decade of advances, multiple topics related to that are exciting.

In fact, the expansion of applications for modulating Nrf2 signaling triggered multiple theoretical studies for understanding the possibilities of designing molecules for enhancing or disrupting Nrf2 actions. Thus, numerous compounds have been proposed as ligands for Nrf2 or proteins involved in its function; among them are those isolated from natural sources, but also some compounds designed to disrupt the interaction of Nrf2 and other proteins. The interactions have been observed from some crystallized structures, but also theoretical approaches have yielded advances to explain the action of compounds active in the Nrf2 signaling pathways, such as bardoxolone, sulforaphane, and resveratrol. The QSAR studies let one identify the key moieties for a specific action, while MD assays show the mechanism for modulating the natural dynamics of Nrf2, or the related proteins in its function. Current trends suggest prompt advances regarding expanding applications of Nrf2 regulation as tools for treating high-impact maladies such as neurodegenerative diseases and metabolism disruption, expanding the studies and applications beyond the treatment of neoplasia in humans.

Acknowledgements

The authors thank Secretaría de Investigación y Posgrado del Instituto Politécnico Nacional for supporting projects in this field.

Conflict of interest

The authors declare no conflict of interest.

Author details

Marvin A. Soriano-Ursúa*, Ana L. Ocampo-Néstor, Melvin N. Rosalez, Eunice D. Farfán-García, Eduardo Osiris Madrigal-Santillán, Ángel Morales-González and José Antonio Morales-González*
Superior School of Medicine, National Polytechnic Institute, Mexico City, Mexico

*Address all correspondence to: msoriano@ipn.mx and jmorales101@yahoo.com.mx

IntechOpen

© 2024 The Author(s). Licensee IntechOpen. This chapter is distributed under the terms of the Creative Commons Attribution License (http://creativecommons.org/licenses/by/3.0), which permits unrestricted use, distribution, and reproduction in any medium, provided the original work is properly cited.

References

[1] Dodson M, de la Vega MR, Cholanians AB, Schmidlin CJ, Chapman E, Zhang DD. Modulating Nrf2 in disease: Timing is everything. Annual Review of Pharmacology and Toxicology. 2019;**59**:555-575. DOI: 10.1146/annurev-pharmtox-010818-021856

[2] Itoh K, Igarashi K, Hayashi N, Nishizawa M, Yamamoto M. Cloning and characterization of a novel erythroid cell-derived CNC family transcription factor heterodimerizing with the small Maf family proteins. Molecular and Cellular Biology. 1995;**15**:4184-4193. DOI: 10.1128/MCB.15.8.4184

[3] Itoh K, Wakabayashi N, Katoh Y, Ishii T, Igarashi K, et al. Keap1 represses nuclear activation of antioxidant responsive elements by Nrf2 through binding to the amino-terminal Neh2 domain. Genes & Development. 1999;**13**:76-86. DOI: 10.1101/gad.13.1.76

[4] Chan K, Lu R, Chang JC, Kan YW. Nrf2, a member of the NFE2 family of transcription factors, is not essential for murine erythropoiesis, growth, and development. Proceedings of the National Academy of Sciences of the United States of America. 1996;**93**:13943-13948. DOI: 10.1073/pnas.93.24.13943

[5] McMahon M, Thomas N, Itoh K, Yamamoto M, Hayes JD. Redox-regulated turnover of Nrf2 is determined by at least two separate protein domains, the redox-sensitive Neh2 degron and the redox-insensitive Neh6 degron. The Journal of Biological Chemistry. 2004;**279**:31556-31567. DOI: 10.1074/jbc.M403061200

[6] Rada P, Rojo AI, Chowdhry S, McMahon M, Hayes JD, Cuadrado A. SCF/{beta}-TrCP promotes glycogen synthase kinase 3-dependent degradation of the Nrf2 transcription factor in a Keap1-independent manner. Molecular and Cellular Biology. 2011;**31**:1121-1133. DOI: 10.1128/MCB.01204-10

[7] Singh A, Misra V, Thimmulappa RK, Lee H, Ames S, et al. Dysfunctional KEAP1-Nrf2 interaction in non-small-cell lung cancer. PLoS Medicine. 2006;**3**(10):e420. DOI: 10.1371/journal.pmed.0030420

[8] Komatsu M, Kurokawa H, Waguri S, Taguchi K, Kobayashi A, et al. The selective autophagy substrate p62 activates the stress responsive transcription factor Nrf2 through inactivation of Keap1. Nature Cell Biology. 2010;**12**:213-223. DOI: 10.1038/ncb2021

[9] Lau A, Wang XJ, Zhao F, Villeneuve NF, Wu T, et al. A noncanonical mechanism of Nrf2 activation by autophagy deficiency: Direct interaction between Keap1 and p62. Molecular and Cellular Biology. 2010;**30**:3275-3285. DOI: 10.1128/MCB.00248-10

[10] Gallorini M, Carradori S, Panieri E, Sova M, Saso L. Modulation of Nrf2: Biological dualism in cancer, targets and possible therapeutic applications. Antioxidants & Redox Signaling. 2024;**40**(10-12):636-662. DOI: 10.1089/ars.2022.0213

[11] Ngo V, Duennwald ML. Nrf2 and oxidative stress: A general overview of mechanisms and implications in human disease. Antioxidants. 2022;**11**(12):2345. DOI: 10.3390/antiox11122345

[12] Kim MJ, Jeon JH. Recent advances in understanding Nrf2 agonism and its potential clinical application to

metabolic and inflammatory diseases. International Journal of Molecular Sciences. 2022;**23**(5):2846. DOI: 10.3390/ijms23052846

[13] Cores Á, Piquero M, Villacampa M, León R, Menéndez JC. Nrf2 regulation processes as a source of potential drug targets against neurodegenerative diseases. Biomolecules. 2020;**10**(6):904. DOI: 10.3390/biom10060904

[14] Suzen S, Tucci P, Profumo E, Buttari B, Saso L. A pivotal role of Nrf2 in neurodegenerative disorders: A new way for therapeutic strategies. Pharmaceuticals. 2022;**15**(6):692. DOI: 10.3390/ph15060692

[15] Chu XY, Liu YM, Zhang HY. Activating or inhibiting Nrf2? Trends in Pharmacological Sciences. 2017;**38**(11):953-955. DOI: 10.1016/j.tips.2017.08.002

[16] Robledinos-Antón N, Fernández-Ginés R, Manda G, Cuadrado A. Activators and inhibitors of Nrf2: A review of their potential for clinical development. Oxidative Medicine and Cellular Longevity. 2019;**2019**(14):9372182. DOI: 10.1155/2019/9372182

[17] Cuadrado A, Manda G, Hassan A, Alcaraz MJ, Barbas C, Daiber A, et al. Transcription factor Nrf2 as a therapeutic target for chronic diseases: A systems medicine approach. Pharmacological Reviews. 2018;**70**(2):348-383. DOI: 10.1124/pr.117.014753

[18] Hybertson BM, Gao B, Bose SK, McCord JM. Oxidative stress in health and disease: The therapeutic potential of Nrf2 activation. Molecular Aspects of Medicine. 2011;**32**(4-6):234-246. DOI: 10.1016/j.mam.2011.10.006

[19] Johnson DA, Johnson JA. Nrf2—A therapeutic target for the treatment of neurodegenerative diseases. Free Radical Biology and Medicine. 2015;**88**:253-267. DOI: 10.1016/j.freeradbiomed.2015.07.147

[20] Beamer LJ, Li X, Bottoms CA, Hannink M. Conserved solvent and side-chain interactions in the 1.35 Å structure of the Kelch domain of Keap1. Acta Crystallographica Section D: Biological Crystallography. 2005;**61**(10):1335-1342. DOI: 10.1107/S0907444905022626

[21] Padmanabhan B, Tong KI, Ohta T, Nakamura Y, Scharlock M, Ohtsuji M, et al. Structural basis for defects of Keap1 activity provoked by its point mutations in lung cancer. Molecular Cell. 2006;**21**(5):689-700. DOI: 10.1016/j.molcel.2006.01.013

[22] Tong KI, Padmanabhan B, Kobayashi A, Shang C, Hirotsu Y, Yokoyama S, et al. Different electrostatic potentials define ETGE and DLG motifs as hinge and latch in oxidative stress response. Molecular and Cellular Biology. 2007;**27**(21):7511-7521. DOI: 10.1128/MCB.00753-07

[23] Williamson TP, Amirahmadi S, Joshi G, Kaludov NK, Martinov MN, Johnson DA, et al. Discovery of potent, novel Nrf2 inducers via quantum modeling, virtual screening, and in vitro experimental validation. Chemical Biology & Drug Design. 2012;**80**(6):810-820. DOI: 10.1111/cbdd.12040

[24] Jnoff E, Albrecht C, Barker JJ, Barker O, Beaumont E, Bromidge S, et al. Binding mode and structure–activity relationships around direct inhibitors of the Nrf2–Keap1 complex. ChemMedChem. 2014;**9**(4):699-705. DOI: 10.1002/cmdc.201300525

[25] Meng XY, Zhang HX, Mezei M, Cui M. Molecular docking: A powerful approach for structure-based drug discovery. Current Computer-Aided

Drug Design. 2011;**7**(2):146-157. DOI: 10.2174/157340911795677602

[26] Agu PC, Afiukwa CA, Orji OU, Ezeh EM, Ofoke IH, Ogbu CO, et al. Molecular docking as a tool for the discovery of molecular targets of nutraceuticals in diseases management. Scientific Reports. 2023;**13**:13398. DOI: 10.1038/s41598-023-40160-2

[27] Ferreira LG, Dos Santos RN, Oliva G, Andricopulo AD. Molecular docking and structure-based drug design strategies. Molecules. 2015;**20**(7):13384-13421. DOI: 10.3390/molecules200713384

[28] Lee S, Hu L. Nrf2 activation through the inhibition of keap1-Nrf2 protein-protein interaction. Medicinal Chemistry Research. 2020;**29**(5):846-867. DOI: 10.1007/s00044-020-02539-y

[29] Crisman E, Duarte P, Dauden E, Cuadrado A, Rodriguez-Franco MI, Lopez MG, et al. KEAP1-Nrf2 protein-protein interaction inhibitors: Design, pharmacological properties and therapeutic potential. Medicinal Research Reviews. 2023;**43**(1):237-287. DOI: 10.1002/med.21925

[30] He F, Ru X, Wen T. Nrf2, a transcription factor for stress response and beyond. International Journal of Molecular Sciences. 2023;**21**:4777. DOI: 10.3390/ijms21134777

[31] Canning P, Sorell FJ, Bullock AN. Structural basis of Keap1 interactions with Nrf2. Free Radical Biology & Medicine. 2015;**88**:101-107. DOI: 10.1016/j.freeradbiomed.2015.05.034

[32] Li M, Huang W, Jie F, Wang M, Zhong Y, Chen Q, et al. Discovery of Keap1-Nrf2 small-molecule inhibitors from phytochemicals based on molecular docking. Food and Chemical Toxicology. 2019;**133**:110758. DOI: 10.1016/j.fct.2019.110758

[33] Guan T, Bian C, Ma Z. In vitro and in silico perspectives on the activation of antioxidant responsive element by citrus-derived flavonoids. Frontiers in Nutrition. 2023;**10**:1257172. DOI: 10.3389/fnut.2023.1257172

[34] Abed DA, Lee S, Hu L. Discovery of disubstituted xylylene derivatives as small molecule direct inhibitors of Keap1-Nrf2 protein-protein interaction. Bioorganic & Medicinal Chemistry. 2020;**28**:115343. DOI: 10.1016/j.bmc.2020.115343

[35] Begnini F, Geschwindner S, Johansson P, Wissler L, Lewis RJ, Danelius E, et al. Importance of binding site hydration and flexibility revealed when optimizing a macrocyclic inhibitor of the Keap1-Nrf2 protein-protein interaction. Journal of Medicinal Chemistry. 2022;**65**:3473-3517. DOI: 10.1021/acs.jmedchem.1c01975

[36] Bello M, Morales-González JA. Molecular recognition between potential natural inhibitors of the Keap1-Nrf2 complex. International Journal of Biological Macromolecules. 2017;**105**:981-992. DOI: 10.1016/j.ijbiomac.2017.07.117

[37] Mou Y, Wen S, Li YX, Gao XX, Zhang X, Jiang ZY. Recent progress in Keap1-Nrf2 protein-protein interaction inhibitors. European Journal of Medicinal Chemistry. 2020;**202**:112532. DOI: 10.1016/j.ejmech.2020.112532

[38] Shimizu Y, Yonezawa T, Sakamolto J, Furuya T, Osawa M, Ikeda K. Identification of novel inhibitors of Keap1/Nrf2 by promising method combining protein-protein interaction-oriented library and machine learning. Scientific Reports. 2021;**11**:7420. DOI: 10.1038/s41598-021-86616-1

[39] Unni S, Deshmukh P, Krishnappa G, Bharath MS, Padmanabhan B. Chlorhexidine as a Keap1-Nrf2 inhibitor: A new target for an old drug for Parkinson's disease therapy. Journal of Biomolecular Structure and Dynamics. 2023;**41**(12):5367-5381. DOI: 10.1080/07391102.2022.2086175

[40] Carrow KP, Hamilton HL, Hopps MP, Li Y, Qiao B, Payne NC, et al. Inhibiting the Keap1/Nrf2 protein-protein interaction with protein-like polymers. Advanced Materials. 2024;**36**(21):e2311467. DOI: 10.1002/adma.202311467

[41] Neves BJ, Braga RC, Melo-Filho CC, Moreira-Filho JT, Muratov EN, Andrade CH. QSAR-based virtual screening: Advances and applications in drug discovery. Frontiers in Pharmacology. 2018;**9**:1275. DOI: 10.3389/fphar.2018.01275

[42] Du A, Jia W. Virtual screening, identification, and potential antioxidant mechanism of novel bioactive peptides during aging by a short-chain peptidomics, quantitative structure-activity relationship analysis, and molecular docking. Food Research International. 2023;**172**:113129. DOI: 10.1016/j.foodres.2023.113129

[43] Liu H, Huang H, Xiao X, Zhao Z, Liu C. Effects of phthalate esters (PAEs) on cell viability and Nrf2 of HepG2 and 3D-QSAR studies. Toxics. 2021;**9**(6):134. DOI: 10.3390/toxics9060134

[44] Kimber I. The activity of methacrylate esters in skin sensitisation test methods II. A review of complementary and additional analyses. Regulatory Toxicology and Pharmacology. 2021;**119**:104821. DOI: 10.1016/j.yrtph.2020.104821

[45] He W, Wang J, Jin Q, Zhang J, Liu Y, Jin Z, et al. Design, green synthesis, antioxidant activity screening, and evaluation of protective effect on cerebral ischemia reperfusion injury of novel monoenone monocarbonyl curcumin analogs. Bioorganic Chemistry. 2021;**114**:105080. DOI: 10.1016/j.bioorg.2021.105080

[46] Dos Santos MM, Rodrigues GCS, de Sousa NF, Scotti MT, Scotti L, Mendonça-Junior FJB. Identification of new targets and the virtual screening of lignans against Alzheimer's disease. Oxidative Medicine and Cellular Longevity. 2020;**2020**(15):3098673. DOI: 10.1155/2020/3098673

[47] Tong JB, Wang J, Luo D, Xiao XC, Xu HY, Bian S, et al. QSAR study, molecular docking, and ADMET prediction of vinyl sulfone-containing Nrf2 activator derivatives for treating Parkinson disease. Structural Chemistry. 2022;**33**(4):1109-1131. DOI: 10.1007/s11224-022-01909-z

[48] Zou L, Yu X, Cai K, Xu B, Chen C, Xiao G. Identification of antioxidant peptides targeting Keap1– Nrf2– ARE pathway from in vitro digestion of pork sausage with partial substitution of NaCl by KCl. Food Research International. 2023;**174**:113585. DOI: 10.1016/j.foodres.2023.113585

[49] Zheng L, Zhao Y, Dong H, Su G, Zhao M. Structure–activity relationship of antioxidant dipeptides: Dominant role of Tyr, Trp, Cys and met residues. Journal of Functional Foods. 2016;**21**:485-496. DOI: 10.1016/j.jff.2015.12.003

[50] Yan W, Lin G, Zhang R, Liang Z, Wu W. Studies on the bioactivities and molecular mechanism of antioxidant peptides by 3D-QSAR, in vitro evaluation and molecular dynamic simulations. Food & Function. 2020;**11**(4):3043-3052. DOI: 10.1039/C9FO03018B

[51] Zhang S, Khan WA, Su L, Zhang X, Li C, Qin W, et al. Predicting oxidative stress induced by organic chemicals by using quantitative structure–activity relationship methods. Ecotoxicology and Environmental Safety. 2020;**201**:110817. DOI: 10.1016/j.ecoenv.2020.110817

[52] Simov V, Altman MD, Bianchi E, DelRizzo S, DiNunzio EN, Feng G, et al. Discovery and characterization of novel peptide inhibitors of the Nrf2/MAFG/DNA ternary complex for the treatment of cancer. European Journal of Medicinal Chemistry. 2021;**224**:113686. DOI: 10.1016/j.ejmech.2021.113686

[53] Chen H, Lou Y, Lin S, Tan X, Zheng Y, Yu H, et al. Formononetin, a bioactive isoflavonoid constituent from Astragalus membranaceus (Fisch.) Bunge, ameliorates type 1 diabetes mellitus via activation of Keap1/Nrf2 signaling pathway: An integrated study supported by network pharmacology and experimental validation. Journal of Ethnopharmacology. 2024;**322**:117576. DOI: 10.1016/j.jep.2023.117576

[54] Raghunath A, Nagarajan R, Sundarraj K, Palanisamy K, Perumal E. Identification of compounds that inhibit the binding of Keap1a/Keap1b Kelch DGR domain with Nrf2 ETGE/DLG motifs in zebrafish. Basic & Clinical Pharmacology & Toxicology. 2019;**125**(3):259-270. DOI: 10.1111/bcpt.13222

[55] Mili A, Das S, Nandakumar K, Lobo R. Molecular docking and dynamics guided approach to identify potential anti-inflammatory molecules as Nrf2 activator to protect against drug-induced liver injury (DILI): A computational study. Journal of Biomolecular Structure and Dynamics. 2023;**41**(19):9193-9210. DOI: 10.1080/07391102.2022.2141885

[56] Wang MQ, Zhang KH, Liu FL, Zhou R, Zeng Y, Chen AL, et al. Wedelolactone alleviates cholestatic liver injury by regulating FXR-bile acid-NF-κB/Nrf2 axis to reduce bile acid accumulation and its subsequent inflammation and oxidative stress. Phytomedicine. 2024;**122**:155124. DOI: 10.1016/j.phymed.2023.155124

[57] Zeng J, Weng Y, Lai T, Chen L, Li Y, Huang Q, et al. Procyanidin alleviates ferroptosis and inflammation of LPS-induced RAW264.7 cell via the Nrf2/HO-1 pathway. Naunyn-Schmiedeberg's Archives of Pharmacology. 2024;**397**(6):4055-4067. DOI: 10.1007/s00210-023-02854-2

[58] Beeraka NM, Zhang J, Zhao D, Liu J, Au C, Vikram PH, et al. Combinatorial implications of Nrf2 inhibitors with FN3K inhibitor: In vitro breast cancer study. Current Pharmaceutical Design. 2023;**29**(30):2408-2425. DOI: 10.2174/0113816128261466231011114600

[59] Fu W, Xiao Z, Chen Y, Pei J, Sun Y, Zhang Z, et al. Molecular integrative study on interaction domains of nuclear factor erythroid 2-related factor 2 with sirtuin 6. Biochimie. 2023;**211**:68-77. DOI: 10.1016/j.biochi.2023.03.002

[60] Dahab M, Zhang P, Al-Mijalli SH, Abdallah EM. Unveiling the anti-cholera and active diabetic Renoprotective compounds of Maqian essential oil: A computational and molecular dynamics study. Molecules. 2023;**28**(24):7954. DOI: 10.3390/molecules28247954

[61] Bello M. Structural basis of Nrf2 activation by flavonolignans from silymarin. Journal of Molecular Graphics and Modelling. 2023;**119**:108393. DOI: 10.1016/j.jmgm.2022.108393

[62] Chang M, Wilson CJ, Karunatilleke NC, Moselhy MH, Karttunen M, Choy WY. Exploring

the conformational landscape of the Neh4 and Neh5 domains of Nrf2 using two different force fields and circular dichroism. Journal of Chemical Theory and Computation. 2021;**17**(5):3145-3156. DOI: 10.1021/acs.jctc.0c01243

[63] Karunatilleke NC, Fast CS, Ngo V, Brickenden A, Duennwald ML, Konermann L, et al. Nrf2, the major regulator of the cellular oxidative stress response, is partially disordered. International Journal of Molecular Sciences. 2021;**22**(14):7434. DOI: 10.3390/ijms22147434

Chapter 3

The Role of NRF2 Transcription Factor in Metabolic Syndrome

Danyelly Bruneska Gondim Martins,
Thaysa Walleria Aragão Santos,
Maria Helena Menezes Estevam Alves
and Rosângela Ferreira Frade de Araújo

Abstract

Metabolic syndrome includes several diseases that are associated with metabolic abnormalities such as obesity, dyslipidemia, hypertension, type 2 diabetes, obesity, cardiac diseases, and insulin resistance. In order to maintain cellular homeostasis, it is necessary to regulate the signaling pathways involved in controlling oxidative stress. Nuclear factor erythroid-2 factor 2 (NRF2) is a transcription factor largely expressed in several tissues and cells and participates in the oxidative stress regulation signaling pathways. NRF2 also mediates transcriptional regulation of a variety of target genes to signalize and regulate acute and chronic stress pathways in metabolic syndrome. Deregulation of NRF2 could contribute to a worst prognosis/profile of individuals with metabolic syndrome. Therefore, NRF2 and its activators might play a role in its treatment, highlighted as targets for modulation by pharmacological agents.

Keywords: oxidative stress, obesity, inflammation, cardiovascular diseases, hypertension

1. Introduction

Metabolic syndrome (MetS) is associated with an increased risk of developing cardiovascular diseases (CVDs) and type 2 diabetes (T2D) and is characterized by several metabolic changes, among which the following stand out: obesity, insulin resistance, hypertension, high levels of triglycerides and fasting glucose, and decreased levels of high-density cholesterol (HDL-c) [1, 2]. The MetS global prevalence can vary from 12.5 to 31.4% being higher in the Eastern Mediterranean Region and Americas [3]. MetS was also associated with a higher risk of incidence of cardiovascular mortality in prospective analyses, increasing for each additional characteristic of MetS observed in the patient [4]. Cardiovascular diseases account for most noncommunicable diseases (NCD) deaths, about 17.9 million people annually, followed by cancers (9.3 million), chronic respiratory diseases (4.1 million), and diabetes (2.0 million including kidney disease deaths caused by diabetes) [5].

Inflammation and oxidative stress are precursors to several complications associated with the findings of MetS, mainly due to increased adiposity, insulin resistance, hypertension, and hyperlipidemia. Oxidative stress is an imbalance between oxidant and antioxidant molecules, leading to changes in signaling pathways, control of redox processes, and cellular damage [6]. The imbalance between prooxidant and antioxidant factors in the body can lead to an increased production of reactive oxygen species (ROS) and nitrogen species (RNS) [7]. ROS play important signaling roles under physiological conditions, once cells deliberately generate ROS through the respiratory chain and other metabolic processes, activating molecular signaling and immune defense mechanisms [8]. Also, ROS regulate several cellular functions such as cell growth, contraction, dilation, migration of vascular cells, and muscle tone [9]. Due to their importance in various physiological functions, certain ROS levels have beneficial effects such as in the process of glucose metabolism, adipogenesis, and adipocyte differentiation [10]. However, increased ROS leads to mitochondrial dysfunction, lipid peroxidation, protein, and DNA damage, compromising antioxidation pathways in metabolic syndrome [11]. Nuclear factor erythroid 2-related factor 2 (NRF2) is known as a master regulator of transcriptional activation of components of antioxidant systems. NRF2 transduces chemical signals to regulate a battery of cytoprotective genes [12]. Target genes include superoxide dismutase (SOD), catalase, glutathione S-transferase (GST), nicotinamide adenine dinucleotide phosphate (NADPH)-oxidase, quinone oxidoreductase 1 (NQO1), γ-glutamate cysteine ligase (GCL), and heme oxygenase-1 (HO-1) [10].

Changes in metabolism and gene expression of redox homeostasis activate NRF2 and other stress response pathways. Furthermore, NRF2 has been identified as a regulator of drug or xenobiotic detoxification, target genes involved in proteasomal and autophagic function, iron, lipid, and carbohydrate metabolism, and DNA repair [13]. In homeostatic conditions, NRF2 binds to its main inhibitor, Kelch-like ECH-associated protein 1 (Keap1), with two specific coupling sites form, a cyclic E3-ubiquitin ligase and Cullin-3/Rbx1 complex. So, NRF2 goes into the ubiquitin proteasome to be degraded. In addition to being an NRF2 inhibitor, Keap 1 acts as a cysteine thiol-rich redox sensor [12]. In the process of oxidative stress, oxidative factors can act and alter the chemical conformation of Keap1 cysteine, thus inhibiting ubiquitination and leading to excessive activation of NRF2 [14]. NRF2 can be phosphorylated by several enzymes, such as protein kinase C (PKC), which phosphorylates it at the Ser40 position, disrupting its interaction with Keap1. ROS can also increase NRF2 translocation to the nucleus, through promotion of the oxidation of cysteine residues in Keap1, allowing the release of NRF2, which will initiate the antioxidant transcription process [15].

Several studies correlate NRF2 with various regulatory functions of genes involved in cell proliferation and differentiation, inflammatory response, and lipid metabolism [12]. It has already confirmed the association between the etiology of MetS, oxidative stress, and chronic inflammation, which are under important control of NRF2 [11]. These data show the importance of investigating the role of NRF2 in this context, which involves changes in metabolic pathways in different organs.

2. NRF2 and metabolic syndrome

The involvement of NRF2 in the development of MetS is based on the interaction of NRF2 with other transcription factors or with the regulation of the expression of molecules that play a role in metabolism. Therefore, the impacts of Nrf2 on MetS require a systemic influence, rather than being controlled by a specific organ. Alterations in NRF2 have been shown to impair adipogenesis through peroxisome proliferator-activated receptor (PPAR), a master regulator of adipocyte differentiation and adipogenesis involved in MetS-related diseases, such as obesity and dyslipidemia [16]. It is known that oxidative stress plays a significant role in inducing insulin resistance, especially in association with the intake of high-fat diets (HFD) and obesity. On the other hand, in pancreatic β cells, excessive oxidative stress suppresses the transcription of insulin-related genes. Once NRF2 regulates redox homeostasis, many studies have been focused on deletion of the NRF2 gene, as well as its overexpression and its pharmacological activators [10].

Liu et al. [17] investigated about the role of NRF2 in nonalcoholic fatty liver disease, a hepatic manifestation of MetS, and suggested that deletion of NRF2 in mice may lead to hepatic insulin resistance due to activation of nuclear factor kappa B (NF-κB), a molecule involved in the phosphorylation of insulin receptor substrates. Consequently, NRF2 activation mitigated the progression of the disease and proved to be a therapeutic strategy. A significant decrease in mitochondrial fatty acid oxidation was observed in the absence of NRF2 and accelerated when NRF2 was constitutively active. These findings have implications for chronic disease conditions, including MetS [18].

2.1 Obesity and dyslipidemia

Commonly recognized for its energy storage function, adipose tissue is also responsible for endocrine functions. Adipocytes can release active biological compounds (adipokines), such as leptin, which act in the control of body weight and play inflammatory functions. This tissue also has different types of cells, such as immune and endothelial cells, which produce different types of hormones, with different functions such as glucose control and lipid metabolism, thus acting as a regulator of metabolic homeostasis [19].

Adipose tissue serves as a pivotal factor in the generation of systemic oxidative stress. The oxidation of fatty acids, within mitochondria and peroxisomes, stands as a significant contributor to oxidative stress, catalyzing the production of ROS. The overconsumption of oxygen also triggers the generation of free radicals within the mitochondrial respiratory chain, intrinsically linked with oxidative phosphorylation [20]. So, it is clear that deregulation in the adipose metabolism can lead to ROS impairment and break down the homeostasis.

Obesity is correlated with low-grade chronic and systemic inflammatory processes. Different studies associate these inflammation mechanisms with severe damage to various organs and tissues and are directly linked to several metabolic diseases [21]. Obese individuals or those with marked weight gain present a phenotypic change in white adipose tissue (WAT) with the appearance of inflamed and dysfunctional adipocytes that secrete pro-inflammatory cytokines [21–23]. These individuals may also have dyslipidemia, which is characterized by a lipid disorder changes in plasma lipoproteins. These alterations in the lipid profile include reduced levels of high-density lipoprotein (HDL) cholesterol, elevated small dense low-density lipoprotein (LDL), and hypertriglyceridemia [24].

Nonalcoholic fatty liver disease (NAFLD), associated with obesity and T2D, is characterized by liver damage caused by a dysfunctional diet, associated with the excess triglycerides in hepatocytes. So, in obese individuals, factors, such as NAFLD, induce oxidative stress in hepatocytes, thus increasing cellular damage.

In addition, hepatic oxidative stress was attenuated with an antioxidant diet in rats with NAFLD, through modulation of the KEAP1/NRF2 pathway [14, 25]. Ma et al. [26] noted that wine devoid of sulfur dioxide effectively alleviated dyslipidemia induced by a HFD, influencing transcription factors, such as NRF2. Then, NRF2 activation enhances lipid metabolism and fatty acid β-oxidation by upregulating crucial genes, such as carnitine palmitoyltransferase I and PARP alpha. Ginsenoside Rg1 (antioxidant molecule extracted from ginseng) significantly improved chronic hepatic inflammatory injuries and fibrosis induced by lipopolysaccharides. This process involves the dissociation of NRF2 from Keap1, subsequently activating NRF2 pathway, which further inhibits NLRP3, NLRP1, and AIM2 inflammasomes. It is important to highlight that the activation of inflammasomes can be mediated by an increase in ROS, contributing to the inflammatory process also present in CVDs [27].

2.2 Hypertension

Reactive oxygen species and changes in radical production can break down redox-sensitive signaling and play an important role in the development of cardiovascular diseases and hypertension, through alterations in the renin-angiotensin system (RAS) and the increased RAS activation and Ang II production lead to an imbalance of redox pathways and to the activation of ROS-producing sources [28].

The harmful effects of the oxidative stress manifest themselves in tissues, inducing endothelial injuries and promoting the enlargement and thickening of the heart walls, especially in the left ventricle. This phenomenon contributes to the development of hypertension and other cardiovascular diseases (CVD) [29]. Increased peripheral resistance, a hallmark of arterial hypertension, arises mainly from structural and functional changes in small resistance arteries, manifested through vascular remodeling and endothelial dysfunction. These changes can result in narrowing the lumen due to a reduced outer diameter (eutrophic remodeling) or an invasion of the lumen by a thicker layer (hypertrophic remodeling) [30].

The interconnection between endothelial function and vascular remodeling is primarily attributed to the decreased availability of nitric oxide (NO), resulting in the promotion of predominantly hypertrophic vascular remodeling [31]. NO is an important vasodilator produced in the endothelium by the catalytic conversion of L-arginine into L-citrulline, through the enzyme endothelial nitric oxide synthase (eNOS). A series of cofactors are involved in this reaction, and their disruption leads to a monomeric form of the enzyme, which is uncoupled, and, instead of NO, superoxide is formed. In both hypertension and atherosclerosis, this uncoupled form of eNOS has been found. Increasingly, evidence indicates that NRF2 activation can trigger vasoprotective effects, reducing the bioavailability of ROS and increasing NO levels by eNOS [15]. In a study using rat pheochromocytoma cells, it was observed that NO can activate NRF2 by S-nitrosylation of Keap1 and, alternatively, by PKC-catalyzed phosphorylation of NRF2 [32]. Therefore, downregulation of NRF2 in hypertension may contribute to decreased antioxidant capacity, resulting in increased oxidative stress and consequent vascular dysfunction.

2.3 Diabetes

Oxidative stress is also causal in the development of β cell dysfunction, diabetes and insulin resistance [33, 34]. Diabetic patients have been shown increased cellular levels of ROS and ROS-induced DNA damage through a variety of mechanisms.

An increased glycemic load in diabetes overwhelms the Krebs cycle, resulting in the inhibition of electron transfer within the mitochondrial membrane. Then, the accumulation of free radicals and ROS could lead to diabetic alterations such as the depletion of natural antioxidant molecules and damage to vascular cells, as well as alterations in gene and protein expression, blood flow, and endothelial cell permeability [33, 35, 36].

The excess of glucose is related to ROS dependent and alterations in NRF2 concentration are involved in the diabetes physiopathology and its complications. In both mouse placental tissue and human placenta cells, both increase and decrease of NRF2 were linked to gestational diabetes and implications for the health of the mother and offspring [37]. Proteins involved in oxidative stress, NRF2-linked phase II enzyme detoxification, and glutathione synthesis showed alterations in gestational diabetes, leading to increased oxidative stress, mitochondrial superoxide generation, protein carbonylation, and DNA damage [38]. In addition, alterations to NRF2 function have been implicated in the onset and progression of type 1 diabetes, through inhibition of the Keap1/NRF2 signaling pathway [39].

Regarding T2DM, the NRF2 has been shown to play a role in mediating diabetic organ-related complications. An important organ-related complication is the pancreatic β cell dysfunction and death induced by ROS. In mice, the genetic NRF2 induction reduced β cell damage, whereas the genetic NRF2 depletion markedly enhanced β cell damage [40, 41]. In addition to the pancreas, NRF2 has a role in diabetes in adipose tissue, skeletal muscle, heart, liver, and kidneys, with the complication of diabetic nephropathy [42].

The treatment of patients with T2DM is challenging since there is no cure available, but glycemia can be controlled by pharmacological therapy [43]. The NRF2 activation results in the activation of other antioxidant enzymes and cytoprotective genes, making it an attractive therapeutic target for treatment of diabetic complications [44]. Antioxidant therapies are comparatively more effective for treating diabetes from which NRF2 and its related factors are one of the potential activators. Moreover, antioxidant-based treatments serve as potential therapeutics for protection against the enhanced oxidative stress, as well as inflammation in diabetes mellitus [45–49]. The diabetic cardiomyopathy induced by T2DM has two main pathogenesis: the oxidative stress status in cardiomyocytes and lipid toxicity [50]. In the late diabetic cardiomyopathy stage, the Nrf2 is significantly downregulated, and then therapies focused on restoring the expression and activity of NRF2 could prevent diabetes-associated injuries in cardiomyocytes [50–52].

2.4 Coronary artery disease

Coronary artery disease (CAD) presents several common complications such as arrhythmias, acute coronary syndrome, congestive heart failure, mitral regurgitation, pericarditis, aneurysm formation, and mural thrombi. Recent research has challenged the protective effects of HDL due to difficulties in raising serum levels and is now focusing on triglyceride-rich lipoproteins (TGRL), in addition to LDL, as causes of atherosclerosis. Traditional risk factors for coronary artery disease have been well combated over the years, such as smoking, LDL levels, and high blood pressure, for example, due to population guidelines and low-cost therapies. However, TGRL are examples of emerging targets as they have greater correlations with the inflammatory state than LDL itself. TGRL cause inflammation, considering, for example, the content of one of its components, apolipoprotein C-III [53, 54].

The pathophysiology of CAD involves both the inflammatory response of the vascular endothelium and oxidative stress. Activation of the NRF2 pathway is associated with a reduction in myocardial infarct size and a delay in the progression to heart failure. The use of NRF2 activation also demonstrated protective effects against ischemia/reperfusion injury. Oxidative stress and inflammation are closely interconnected processes with oxidative stress-induced inflammation predominantly driven by NF-κB activation. It plays a crucial role in regulating the inflammatory response by controlling the gene expression of inflammatory cytokines, which are the main contributors to cardiovascular diseases in humans [54, 55].

Several mechanisms lead to the complex connection between the NRF2 and NF-kB pathways, which occur by inhibiting each other. IκB-α is an inhibitor of the NF-κB transcription factor as it prevents, for example, the binding of NF-kB to DNA, making its action impossible. NRF2 prevents proteasomal degradation of IκB-α and inhibits nuclear translocation of NF-κB. This occurs due to the increase in HO-1 levels stimulated by NRF2 and subsequent increase in the expression of phase II enzymes that block IκB degradation. NRF2 acts to decrease the expression of pro-inflammatory cytokines such as vascular cell adhesion molecule-1 (VCAM-1), monocyte chemoattractant protein-1 (MCP-1), and tumor necrosis factor-alpha (TNF-α), interferon-gamma (IFN-γ), and some interleukins, all involved in atherosclerosis. Understanding this intricate interaction is essential to unravel the inflammatory mechanisms involved in the pathogenesis of CAD and develop more specific therapeutic interventions [56–58].

3. NRF2 and therapy

Most pharmacological activators of NRF2 are electrophilic substances capable of modifying cysteine residues present in Keap1, ultimately resulting in changes in its conformation through oxidation or alkylation processes [15]. Some of these new therapies and medications have been considered to control MetS through regulation of Keap1 protein and NRF2 (**Figure 1**).

Cannabidiol (CBD) is a nontoxic compound derived from cannabis, known for its anti-inflammatory, anti-apoptotic, and antioxidant properties. Interest in the use of CBD has grown, including studies focusing on cardiovascular diseases. In mice, it was observed increased translocation of NRF2 from the cytoplasm to the nucleus *via* Keap1 inhibition, recognized as a beneficial effect of CBD, preventing acute myocardial infarction after exhaustive exercise [59].

In an environment of oxidative stress NRF2 is released from Keap1, allowing the nuclear translocation of NRF2 and the increase of expression of antioxidant proteins. The action of drugs on MetS components (obesity, diabetes, hypertension, and coronary artery disease) may lead to the decrease of Keap1, thus increasing the activation of NRF2 and HO-1, NRF2 nuclear translocation, and activation of antioxidant protein expression. The drugs also act to reduce inflammatory cytokines and NF-Kb, which in normal oxidative conditions act to inhibit NRF2, impairing translocation.

Resveratrol, administered to hypertensive rats, improved NRF2 activity by combating oxidative stress in proximal tubular epithelial cells, improving inflammation, and attenuating the development of hypertension [60]. Lopes et al. [30] observed a vasoprotective function of NRF2 in hypertension when evaluating hypertensive rats under conditions of vascular oxidative stress related to vascular dysfunction. Bardoxolone and L-sulforaphane activate NRF2 when administered to animals,

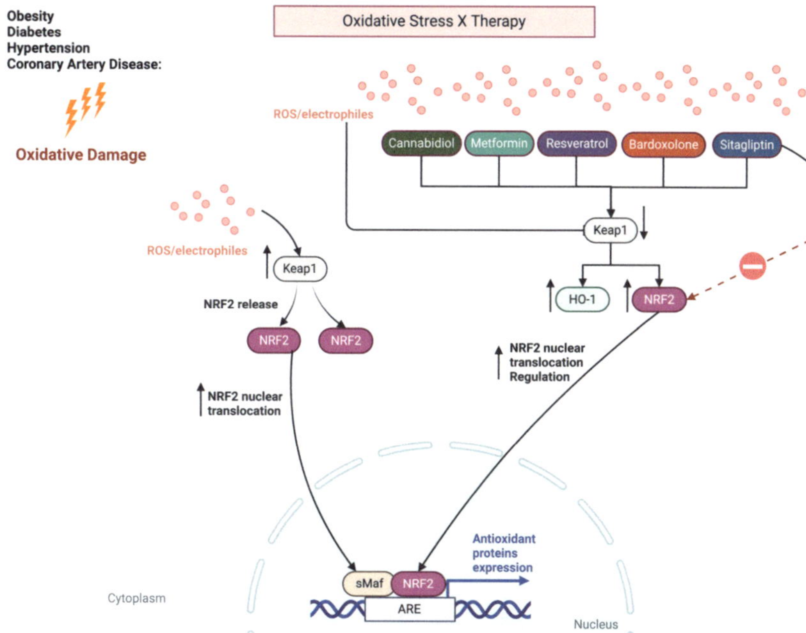

Figure 1.
Effect of oxidative stress and antioxidant drugs used in the treatment of metabolic syndrome.

reducing the formation of ROS that are induced by angiotensin II, and reestablishing the endothelial function in the cells. The impact of oxidative stress, induced by ANG II infusion, was also investigated by Wang et al. [61], in mice, analyzing the effects of treatment with tert-butylhydroquinone (tBHQ), that triggered the activation of NRF2, preventing increased oxidative stress, microvascular ROS, endothelial dysfunction, enhanced microvascular remodeling, contractility, and hypertension. In a study by Gao et al. [62], Qing brick tea (QBT), a traditional and popular Chinese drink, was administered to mice with obesity induced by monosodium glutamate, proving effective in protecting against metabolic syndrome. QBT treatment attenuated oxidative stress and IR and markedly stimulated the NRF2/HO-1 and insulin signaling pathways. A study based on the transcriptome network analysis after using methylglyoxal, conducted on the visceral adipose tissue of hereditary hypertriglyceridemic rats, a nonobese model of metabolic syndrome, revealed that the downregulation of NRF2 expression resulted in the over-representation of genes associated with insulin signaling, lipid metabolism, and angiogenesis [63].

Some drugs already used to control glycemia act regulating the activity of NRF2. Among these, Sitagliptin is a drug indicated for use combined with metformin or sulfonylurea, for treating T2DM. Recent studies indicate that Sitagliptin can also have a regulatory effect on autophagy and protect the cardiovascular system. In the cell, it acts to reduce NF-kB, which, in turn, regulates inflammatory cytokines and helps to inhibit the overexpression of NRF2 [64, 65]. Metformin is the first-line pharmacotherapy for T2DM. Studies carried out with mice to understand the extent of the drug's use have shown that its long-term use not only helps in the treatment of T2DM but also acts as a cardioprotector against heart diseases associated with MetS. These studies indicate that Metformin increases the expression of Nrf2, inducing an improvement in the nuclear translocation of Nrf2 [65, 66]. It reduces excessive

ROS generation, limiting cell apoptosis, and increasing signaling through the ROS-mediated PI3K/Akt and NRF2/HO-1 pathways and the levels of antioxidants [66].

Diabetic nephropathy is an important diabetes complication characterized by inflammatory response and renal dysfunction [48]. Sulforaphane (SFN) is an organosulfur compound that exhibits anticancer and antidiabetic properties and has an indirect antioxidant effect inducing expression of several enzymes *via* the Keap1/NRF2 pathway [67, 68]. Then, SFN has particular interest in the role *via* induction of NRF2 of preventing diabetic nephropathy and renal damage [69]. The upregulation of NRF2 of diabetic nephropathy mice model, by treatment with fenofibrate, led to regulation of other antioxidant molecules and delayed the progression of diabetic nephropathy [46]. Baicalin (BAI), a flavonoid component extracted from the traditional Chinese medicine *S. baicalensis*, showed to inhibit oxidative stress in hyperglycemic environments through regulation of NRF2 pathways. It also decreases blood lipids and inflammation markers and suppresses the excessive production of ROS in cardiovascular diseases [50, 70, 71].

4. Conclusion

Considering the role of NRF2 in activating genes involved in combating oxidative stress and inflammation, it is clear that its inhibition, which often occurs due to increased Keap1 activity, can significantly increase the chances of developing metabolic syndrome. This association is justified by the participation of these genes, activated by this important transcription factor, in several metabolic pathways related especially to endothelial function, insulin signaling, and lipid catabolism. So, it impacts the set of clinical findings that constitute metabolic syndrome. Therefore, several drugs that act on NRF2 activation are promising for improving or preventing chronic metabolic diseases that are not only common but also increasing, assumed as one of the diseases of the century.

Conflict of interest

The authors declare no conflict of interest.

DOI: http://dx.doi.org/10.5772/intechopen.1005035

Author details

Danyelly Bruneska Gondim Martins[1,2]*, Thaysa Walleria Aragão Santos[1], Maria Helena Menezes Estevam Alves[1] and Rosângela Ferreira Frade de Araújo[1,2]

1 Keizo Asami Institute (iLIKA), Federal University of Pernambuco, Brazil

2 Department of Biochemistry, Federal University of Pernambuco, Brazil

*Address all correspondence to: bruneska@prospecmol.org

IntechOpen

© 2024 The Author(s). Licensee IntechOpen. This chapter is distributed under the terms of the Creative Commons Attribution License (http://creativecommons.org/licenses/by/3.0), which permits unrestricted use, distribution, and reproduction in any medium, provided the original work is properly cited.

References

[1] Simmons RK, Alberti KG, Gale EA, Colagiuri S, Tuomilehto J, Qiao Q, et al. The metabolic syndrome: Useful concept or clinical tool? Report of a WHO expert consultation. Diabetologia. 2010;**53**(4):600-605. DOI: 10.1007/s00125-009-1620-4

[2] Saklayen MG. The global epidemic of the metabolic syndrome. Current Hypertension Reports. 2018;**20**(2):12. DOI: 10.1007/s11906-018-0812-z

[3] Noubiap JJ, Nansseu JR, Lontchi-Yimagou E, Nkeck JR, Nyaga UF, Ngouo AT, et al. Geographic distribution of metabolic syndrome and its components in the general adult population: A meta-analysis of global data from 28 million individuals. Diabetes Research and Clinical Practice. 2022;**188**:109924. DOI: 10.1016/j.diabres.2022.109924

[4] Guembe MJ, Fernandez-Lazaro CI, Sayon-Orea C, Toledo E, Moreno-Iribas C, RIVANA Study Investigators. Risk for cardiovascular disease associated with metabolic syndrome and its components: A 13-year prospective study in the RIVANA cohort. Cardiovascular Diabetology. 2020;**19**(1):195. DOI: 10.1186/s12933-020-01166-6

[5] WHO. World Health Organization (WHO). 2023. Available from: https://www.who.int/news-room/fact-sheets/detail/noncommunicable-diseases [Accessed: Jan 20, 2024]

[6] Singh CK, Chhabra G, Ndiaye MA, Garcia-Peterson LM, Mack NJ, Ahmad N. The role of sirtuins in antioxidant and redox signaling. Antioxidants & Redox Signaling. 2018;**28**(8):643-661. DOI: 10.1089/ars.2017.7290

[7] Cadenas S. Mitochondrial uncoupling, ROS generation and cardioprotection. Biochimica et Biophysica Acta - Bioenergetics. 2018;**1859**(9):940-950. DOI: 10.1016/j.bbabio.2018.05.019

[8] Goszcz K, Deakin SJ, Duthie GG, Stewart D, Leslie SJ, Megson IL. Antioxidants in cardiovascular therapy: Panacea or false hope? Frontiers in Cardiovascular Medicine. 2015;**2**:29. DOI: 10.3389/fcvm.2015.00029

[9] Montezano AC, Tsiropoulou S, Dulak-Lis M, Harvey A, Camargo Lde L, Touyz RM. Redox signaling, Nox5 and vascular remodeling in hypertension. Current Opinion in Nephrology and Hypertension. 2015;**24**(5):425-433. DOI: 10.1097/MNH.0000000000000153

[10] Zhang Z, Zhou S, Jiang X, Wang YH, Li F, Wang YG, et al. The role of the Nrf2/Keap1 pathway in obesity and metabolic syndrome. Reviews in Endocrine & Metabolic Disorders. 2015;**16**(1):35-45. DOI: 10.1007/s11154-014-9305-9

[11] Masenga SK, Kabwe LS, Chakulya M, Kirabo A. Mechanisms of oxidative stress in metabolic syndrome. International Journal of Molecular Sciences. 2023;**24**(9):7898. DOI: 10.3390/ijms24097898

[12] Yamamoto M, Kensler TW, Motohashi H. The KEAP1-NRF2 system: A thiol-based sensor-effector apparatus for maintaining redox homeostasis. Physiological Reviews. 2018;**98**(3):1169-1203. DOI: 10.1152/physrev.00023.2017

[13] Dodson M, Castro-Portuguez R, Zhang DD. NRF2 plays a critical role in mitigating lipid peroxidation and ferroptosis. Redox Biology.

2019;**23**:101107. DOI: 10.1016/j.redox.2019.101107. Epub 2019 Jan 11

[14] Lu L, Fu J, Sun J, Liu D, Chen C, Wang H, et al. Is Nrf2-ARE a potential target in NAFLD mitigation? Current Opinion in Toxicology. 2019;**13**:35-44. DOI: 10.1016/j.cotox.2018.12.005

[15] Tanase DM, Apostol AG, Costea CF, Tarniceriu CC, Tudorancea I, Maranduca MA, et al. Oxidative stress in arterial hypertension (HTN): The nuclear factor erythroid factor 2-related factor 2 (Nrf2) pathway, implications and future perspectives. Pharmaceutics. 2022;**14**(3):534. DOI: 10.3390/pharmaceutics14030534

[16] Pi J, Leung L, Xue P, Wang W, Hou Y, Liu D, et al. Deficiency in the nuclear factor E2-related factor-2 transcription factor results in impaired adipogenesis and protects against diet-induced obesity. The Journal of Biological Chemistry. 2010;**285**(12):9292-9300. DOI: 10.1074/jbc.M109.093955

[17] Liu Z, Dou W, Ni Z, Wen Q, Zhang R, Qin M, et al. Deletion of Nrf2 leads to hepatic insulin resistance via the activation of NF-κB in mice fed a high-fat diet. Molecular Medicine Reports. 2016;**14**(2):1323-1331. DOI: 10.3892/mmr.2016.5393

[18] Ludtmann MH, Angelova PR, Zhang Y, Abramov AY, Dinkova-Kostova AT. Nrf2 affects the efficiency of mitochondrial fatty acid oxidation. The Biochemical Journal. 2014;**457**(3):415-424. DOI: 10.1042/BJ20130863

[19] Li S, Eguchi N, Lau H, Ichii H. The role of the Nrf2 signaling in obesity and insulin resistance. International Journal of Molecular Sciences. 2020;**21**(18):6973. DOI: 10.3390/ijms21186973

[20] Fernández-Sánchez A, Madrigal-Santillán E, Bautista M, Esquivel-Soto J, Morales-González A, Esquivel-Chirino C, et al. Inflammation, oxidative stress, and obesity. International Journal of Molecular Sciences. 2011;**12**(5):3117-3132. DOI: 10.3390/ijms12053117

[21] Kawai T, Autieri MV, Scalia R. Adipose tissue inflammation and metabolic dysfunction in obesity. American Journal of Physiology. Cell Physiology. 2021;**320**(3):C375-C391. DOI: 10.1152/ajpcell.00379.2020

[22] Hotamisligil GS. Inflammation and metabolic disorders. Nature. 2006;**444**(7121):860-867. DOI: 10.1038/nature05485

[23] Hotamisligil GS. Inflammation, metaflammation and immunometabolic disorders. Nature. 2017;**542**(7640):177-185. DOI: 10.1038/nature21363

[24] Vekic J, Stefanovic A, Zeljkovic A. Obesity and dyslipidemia: A review of current evidence. Current Obesity Reports. 2023;**12**(3):207-222. DOI: 10.1007/s13679-023-00518-z

[25] Wang Y, Liu B, Wu P, Chu Y, Gui S, Zheng Y, et al. Dietary selenium alleviated mouse liver oxidative stress and NAFLD induced by obesity by regulating the KEAP1/NRF2 pathway. Antioxidants (Basel). 2022;**11**(2):349. DOI: 10.3390/antiox11020349

[26] Ma Y, Yu K, Wang N, Xiao X, Leng Y, Fan J, et al. Sulfur dioxide-free wine with polyphenols promotes lipid metabolism via the Nrf2 pathway and gut microbiota modulation. Food Chemistry: X. 2023;**21**:101079. DOI: 10.1016/j.fochx.2023.101079

[27] Zhou H, Liu Y, Su Y, Ji P, Kong L, Sun R, et al. Ginsenoside Rg1 attenuates

lipopolysaccharide-induced chronic liver damage by activating Nrf2 signaling and inhibiting inflammasomes in hepatic cells. Journal of Ethnopharmacology. 2024;**324**:117794. DOI: 10.1016/j.jep.2024.117794

[28] Dovinova I, Kvandová M, Balis P, Gresova L, Majzunova M, Horakova L, et al. The role of Nrf2 and PPARgamma in the improvement of oxidative stress in hypertension and cardiovascular diseases. Physiological Research. 2020;**69**(Suppl. 4):S541-S553. DOI: 10.33549/physiolres.934612

[29] Liu QQ, Ren K, Liu SH, Li WM, Huang CJ, Yang XH. MicroRNA-140-5p aggravates hypertension and oxidative stress of atherosclerosis via targeting Nrf2 and Sirt2. International Journal of Molecular Medicine. 2019;**43**:839-849. DOI: 10.3892/ijmm.2018.3996

[30] Lopes RA, Neves KB, Tostes RC, Montezano AC, Touyz RM. Downregulation of nuclear factor erythroid 2-related factor and associated antioxidant genes contributes to redox-sensitive vascular dysfunction in hypertension. Hypertension. 2015;**66**(6):1240-1250. DOI: 10.1161/hypertensionaha.115.06163

[31] Masi S, Georgiopoulos G, Chiriacò M, Grassi G, Seravalle G, Savoia C, et al. The importance of endothelial dysfunction in resistance artery remodelling and cardiovascular risk. Cardiovascular Research. 2020;**116**(2):429-437. DOI: 10.1093/cvr/cvz096

[32] Um HC, Jang JH, Kim DH, Lee C, Surh YJ. Nitric oxide activates Nrf2 through S-nitrosylation of Keap1 in PC12 cells. Nitric Oxide. 2011;**25**(2):161-168. DOI: 10.1016/j.niox.2011.06.001

[33] Nowotny K, Jung T, Höhn A, Weber D, Grune T. Advanced glycation end products and oxidative stress in type 2 diabetes mellitus. Biomolecules. 2015;**5**(1):194-222. DOI: 10.3390/biom5010194

[34] Henriksen EJ, Diamond-Stanic MK, Marchionne EM. Oxidative stress and the etiology of insulin resistance and type 2 diabetes. Free Radical Biology & Medicine. 2011;**51**(5):993-999. DOI: 10.1016/j.freeradbiomed.2010.12.005

[35] David JA, Rifkin WJ, Rabbani PS, Ceradini DJ. The Nrf2/Keap1/ARE pathway and oxidative stress as a therapeutic target in type II diabetes mellitus. Journal Diabetes Research. 2017;**2017**:4826724. DOI: 10.1155/2017/4826724

[36] Giacco F, Brownlee M. Oxidative stress and diabetic complications. Circulation Research. 2010;**107**(9):1058-1070. DOI: 10.1161/CIRCRESAHA.110.223545

[37] Zhang C, Yang Y, Chen R, Wei Y, Feng Y, Zheng W, et al. Aberrant expression of oxidative stress related proteins affects the pregnancy outcome of gestational diabetes mellitus patients. American Journal of Translational Research. 2019;**11**(1):269-279

[38] Cheng X, Chapple SJ, Patel B, Puszyk W, Sugden D, Yin X, et al. Gestational diabetes mellitus impairs Nrf2-mediated adaptive antioxidant defenses and redox signaling in fetal endothelial cells in utero. Diabetes. 2013;**62**(12):4088-4097. DOI: 10.2337/db13-0169

[39] Lou Y, Kong M, Li L, Hu Y, Zhai W, Qi X, et al. Inhibition of the Keap1/Nrf2 signaling pathway significantly promotes the progression of type 1 diabetes mellitus. Oxidative Medicine and

Cellular Longevity. 2021;**2021**:7866720. DOI: 10.1155/2021/7866720

[40] Yagishita Y, Fukutomi T, Sugawara A, Kawamura H, Takahashi T, Pi J, et al. Nrf2 protects pancreatic β-cells from oxidative and nitrosative stress in diabetic model mice. Diabetes. 2014;**63**(2):605-618. DOI: 10.2337/db13-0909

[41] Uruno A, Furusawa Y, Yagishita Y, Fukutomi T, Muramatsu H, Negishi T, et al. The Keap1-Nrf2 system prevents onset of diabetes mellitus. Molecular and Cellular Biology. 2013;**33**(15):2996-3010. DOI: 10.1128/MCB.00225-13

[42] Dodson M, Shakya A, Anandhan A, Chen J, Garcia JGN, Zhang DD. NRF2 and diabetes: The good, the bad, and the complex. Diabetes. 2022;**71**(12):2463-2476. DOI: 10.2337/db22-0623

[43] Artasensi A, Pedretti A, Vistoli G, Fumagalli L. Type 2 diabetes mellitus: A review of multi-target drugs. Molecules. 2020;**25**(8):1987. DOI: 10.3390/molecules25081987

[44] Tan SM, de Haan JB. Combating oxidative stress in diabetic complications with Nrf2 activators: How much is too much? Redox Report. 2014;**19**(3):107-117. DOI: 10.1179/1351000214Y.0000000087

[45] Zheng H, Whitman SA, Wu W, Wondrak GT, Wong PK, Fang D, et al. Therapeutic potential of Nrf2 activators in streptozotocin-induced diabetic nephropathy. Diabetes. 2011;**60**(11):3055-3066. DOI: 10.2337/db11-0807

[46] Yang Y, Chen G, Cheng X, Teng Z, Cai X, Yang J, et al. Therapeutic potential of digitoflavone on diabetic nephropathy: Nuclear factor erythroid 2-related factor 2-dependent anti-oxidant and anti-inflammatory effect. Scientific Reports. 2015;**5**:12377. DOI: 10.1038/srep12377

[47] Wu P, Yan Y, Ma LL, Hou BY, He YY, Zhang L, et al. Effects of the Nrf2 protein modulator salvianolic acid a alone or combined with metformin on diabetes-associated macrovascular and renal injury. The Journal of Biological Chemistry. 2016;**291**(42):22288-22301. DOI: 10.1074/jbc.M115.712703

[48] Li S, Zheng L, Zhang J, Liu X, Wu Z. Inhibition of ferroptosis by up-regulating Nrf2 delayed the progression of diabetic nephropathy. Free Radical Biology & Medicine. 2021;**162**:435-449. DOI: 10.1016/j.freeradbiomed.2020.10.323

[49] Dwivedi DK, Jena GB. Dimethyl fumarate-mediated Nrf2/ARE pathway activation and glibenclamide-mediated NLRP3 inflammasome cascade inhibition alleviate type II diabetes-associated fatty liver in rats by mitigating oxidative stress and inflammation. Journal of Biochemical and Molecular Toxicology. 2023;**37**(7):e23357. DOI: 10.1002/jbt.23357

[50] Li R, Liu Y, Shan YG, Gao L, Wang F, Qiu CG. Bailcalin protects against diabetic cardiomyopathy through Keap1/Nrf2/AMPK-mediated antioxidative and lipid-lowering effects. Oxidative Medicine and Cellular Longevity. 2019;**2019**:3206542. DOI: 10.1155/2019/3206542

[51] He X, Kan H, Cai L, Ma Q. Nrf2 is critical in defense against high glucose-induced oxidative damage in cardiomyocytes. Journal of Molecular and Cellular Cardiology. 2009;**46**(1):47-58. DOI: 10.1016/j.yjmcc.2008.10.007

[52] Tan Y, Ichikawa T, Li J, Si Q, Yang H, Chen X, et al. Diabetic downregulation

of Nrf2 activity via ERK contributes to oxidative stress-induced insulin resistance in cardiac cells in vitro and in vivo. Diabetes. 2011;**60**(2):625-633. DOI: 10.2337/db10-1164

[53] Libby P. The changing landscape of atherosclerosis. Nature. 2021;**592**(7855):524-533. DOI: 10.1038/s41586-021-03392-8

[54] Casper E. The crosstalk between Nrf2 and NF-κB pathways in coronary artery disease: Can it be regulated by SIRT6? Life Sciences. 2023;**330**:122007. DOI: 10.1016/j.lfs.2023.122007

[55] Battino M, Giampieri F, Pistollato F, Sureda A, de Oliveira MR, Pittalà V, et al. Nrf2 as regulator of innate immunity: A molecular Swiss army knife! Biotechnology Advances. 2018;**36**(2):358-370. DOI: 10.1016/j.biotechadv.2017.12.012

[56] Saha S, Buttari B, Panieri E, Profumo E, Saso L. An overview of Nrf2 signaling pathway and its role in inflammation. Molecules. 2020;**25**(22):5474. DOI: 10.3390/molecules25225474

[57] Coto E, Reguero JR, Avanzas P, Pascual I, Martín M, Hevia S, et al. Gene variants in the NF-KB pathway (NFKB1, NFKBIA, NFKBIZ) and risk for early-onset coronary artery disease. Immunology Letters. 2019;**208**:39-43. DOI: 10.1016/j.imlet.2019.02.007

[58] Tanriover C, Copur S, Mutlu A, Peltek IB, Galassi A, Ciceri P, et al. Early aging and premature vascular aging in chronic kidney disease. Clinical Kidney Journal. 2023;**16**(11):1751-1765. DOI: 10.1093/ckj/sfad076

[59] Zhang J, Luo Z, Zhang Z, Zhao M, Tong C, Cong P, et al. Protective effect and mechanism of cannabidiol on myocardial injury in exhaustive exercise training mice. Chemico-Biological Interactions. 2022;**365**:110079. DOI: 10.1016/j.cbi.2022.110079

[60] Javkhedkar AA, Quiroz Y, Rodriguez-Iturbe B, Vaziri ND, Lokhandwala MF, Banday AA. Resveratrol restored Nrf2 function, reduced renal inflammation, and mitigated hypertension in spontaneously hypertensive rats. American Journal of Physiology. Regulatory, Integrative and Comparative Physiology. 2015;**308**(10):R840-R846. DOI: 10.1152/ajpregu.00308.2014

[61] Wang C, Luo Z, Carter G, Wellstein A, Jose PA, Tomlinson J, et al. NRF2 prevents hypertension, increased ADMA, microvascular oxidative stress, and dysfunction in mice with two weeks of ANG II infusion. American Journal of Physiology. Regulatory, Integrative and Comparative Physiology. 2018;**314**(3):R399-R406. DOI: 10.1152/ajpregu.00122.2017

[62] Gao W, Xiao C, Hu J, Chen B, Wang C, Cui B, et al. Qing brick tea (QBT) aqueous extract protects monosodium glutamate-induced obese mice against metabolic syndrome and involves up-regulation transcription factor nuclear factor-erythroid 2-related factor 2 (Nrf2) antioxidant pathway. Biomedicine & Pharmacotherapy. 2018;**103**:637-644. DOI: 10.1016/j.biopha.2018.04.043

[63] Hüttl M, Markova I, Miklankova D, Makovicky P, Pelikanova T, Šeda O, et al. Adverse effects of methylglyoxal on transcriptome and metabolic changes in visceral adipose tissue in a Prediabetic rat model. Antioxidants (Basel). 2020;**9**(9):803. DOI: 10.3390/antiox9090803

[64] Gu Y, Ma CT, Gu HL, Shi L, Tian XT, Xu WQ. Sitagliptin improves

cardiac function after myocardial infarction through activation of autophagy in streptozotocin-induced diabetic mice. European Review for Medical and Pharmacological Sciences. 2018;**22**(24):8973-8983. DOI: 10.26355/eurrev_201812_16668

[65] Yi M, Cruz Cisneros L, Cho EJ, Alexander M, Kimelman FA, Swentek L, et al. Nrf2 pathway and oxidative stress as a common target for treatment of diabetes and its comorbidities. International Journal of Molecular Sciences. 2024;**25**:821. DOI: 10.3390/ijms25020821

[66] Wang G, Wang Y, Yang Q, Xu C, Zheng Y, Wang L, et al. Metformin prevents methylglyoxal-induced apoptosis by suppressing oxidative stress in vitro and in vivo. Cell Death & Disease. 2022;**13**(1):29. DOI: 10.1038/s41419-021-04478-x

[67] Clarke JD, Hsu A, Williams DE, Dashwood RH, Stevens JF, Yamamoto M, et al. Metabolism and tissue distribution of sulforaphane in Nrf2 knockout and wild-type mice. Pharmaceutical Research. 2011;**28**(12):3171-3179. DOI: 10.1007/s11095-011-0500-z

[68] Yoon HY, Kang NI, Lee HK, Jang KY, Park JW, Park BH. Sulforaphane protects kidneys against ischemia-reperfusion injury through induction of the Nrf2-dependent phase 2 enzyme. Biochemical Pharmacology. 2008;**75**(11):2214-2223. DOI: 10.1016/j.bcp.2008.02.029

[69] Cui W, Bai Y, Miao X, Luo P, Chen Q, Tan Y, et al. Prevention of diabetic nephropathy by sulforaphane: Possible role of Nrf2 upregulation and activation. Oxidative Medicine and Cellular Longevity. 2012;**2012**:821936. DOI: 10.1155/2012/821936

[70] Hang Y, Qin X, Ren T, Cao J. Baicalin reduces blood lipids and inflammation in patients with coronary artery disease and rheumatoid arthritis: A randomized, double-blind, placebo-controlled trial. Lipids in Health and Disease. 2018;**17**(1):146. DOI: 10.1186/s12944-018-0797-2

[71] Wang G, Liang J, Gao LR, et al. Baicalin administration attenuates hyperglycemia-induced malformation of cardiovascular system. Cell Death & Disease. 2018;**9**(2):234. DOI: 10.1038/s41419-018-0318-2

Chapter 4

The Dual Role of NRF2 Transcription Factor in Female Cancer

Danyelly Bruneska Gondim Martins, Ananda Cristina de Aguiar, Francielle Maria de Araújo Barbosa and Glauber Moreira Leitão

Abstract

Nuclear factor erythroid 2-related factor 2 (NRF2) is an essential transcription factor that is involved in cellular defense against oxidative stress and is assumed to be an important molecule in the transcription and regulation of cytoprotective genes. NRF2 is not only responsible for protecting healthy cells but plays a role in neoplastic cells once high expression of NRF2 has been observed in cancer cells. However, the increase in NRF2 levels may be correlated with resistance to therapy, making it interesting to understand the duality of the protective action of this molecule in the scenario of the cancer hallmarks, NRF2-regulated target genes involved in redox homeostasis, drug metabolism and excretion, amino acid metabolism, iron metabolism, energetic metabolism, survival, autophagy, proliferation, DNA repair, proteasomal degradation, and mitochondrial physiology. Therefore, NRF2 has emerged as a promising target in cancer treatment, and many efforts have been made to identify therapeutic strategies that inhibit its oncogenic role.

Keywords: oxidative stress, inflammation, cancer, chemoresistance, female cancer, cancer therapy

1. Introduction

In aerobic organisms, the primary source of adenosine triphosphate (ATP) is oxidative phosphorylation. During this process, reactive oxygen species (ROS) are regularly produced in the cells as a result of oxygen's role, the last electron acceptor in oxidative phosphorylation. However, excessive ROS generation can damage macromolecules like proteins, lipids, and DNA. Then, the cells respond and adapt to stresses through defined regulatory mechanisms, allowing them to survive and maintain homeostasis in their environment [1].

Nuclear factor erythroid 2-related factor 2 (NRF2) is known as a key regulator of the cellular xenobiotic and oxidative stress responses. It is a member of the Cap'n'Collar (CNC) subfamily of basic leucine zipper (bZip) transcription factors, which is encoded by the nuclear factor erythroid-derived 2-like 2 (NFE2L2) gene [2]. NRF2 contains 605 amino acids that are divided into seven conserved regions known

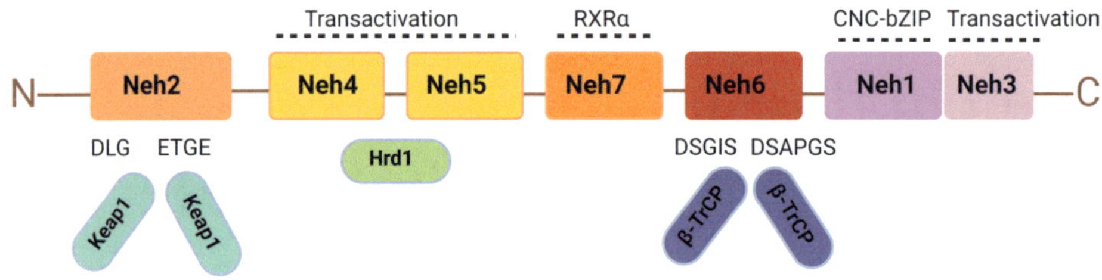

Figure 1.
Structure of Nrf2.

as the Nrf2-ECH homology (Neh) domains with distinct functions (**Figure 1**). The Neh1 domain mediates the dimerization of NRF2 with small musculoaponeurotic fibrosarcomas (MAFs) [3], while Neh2 domain mediates weak and strong interactions with Kelch-like ECH-associated protein 1 (KEAP1) [4]. The C-terminal Neh3 domain acts as a transcriptional activation domain, while Neh4 and the Neh5 domains mediate transactivation or transrepression by coactivators and corepressors [2]. Rich in serine residues, the Neh6 domain has two β-transducin repeat-containing protein (β-TrCP) degrons, DSGIS and DSAPGS, which are involved in the degradation of NRF2 [5]. The Neh7 domain interacts with retinoic X receptor alpha (RXRα), which suppresses NRF2/antioxidant response element (ARE) and inhibits the cytoprotective activity of NRF2 [6].

Itoh et al. [7] found that the Nrf2 functions as an essential transcription factor that interacts with the antioxidant response element (ARE), increasing the transcription of several genes involved in cytoprotection and antioxidation [7]. Nrf2 activity is negatively regulated by KEAP1 [4], so when NRF2 is upregulated due to oxidative or electrophilic alteration of KEAP1 cysteines, it is referred to as canonical NRF2 activation. This modification leads to a conformational change of the Nrf2-Keap1 complex, preventing Nrf2 degradation [8]. The term non-canonical activation of the NRF2 signaling pathway describes the accumulation of NRF2 protein, increased by ubiquitin-binding protein p62's sequestration of KEAP1 during autophagic dysfunction. The p62 can compete with NRF2 for KEAP1 binding, and the p62-KEAP1 complex is then recruited to the autophagosome for degradation [9]. In non-stressed conditions, Nrf2 is constitutively degraded via the proteasome pathway. However, exposure to oxidative stress stimulates Nrf2 activation through the loss of Keap1 affinity. Then, NRF2 can enter the nucleus and attach to the ARE in the promoter region of target genes [10]. Nrf2 activation has shown potential for organ protection against toxins, but overactivation may drive mutations and has been associated with cancer progression and drug resistance, revealing a dual behavior. Nevertheless, the negative regulation of Nrf2 has been assumed as a promising avenue for new cancer therapies [11].

2. NFR2 involved in hallmarks of cancer

The findings on the regulation of NRF2 in the tumor microenvironment are still contradictory. In the tumor microenvironment, NRF2 signaling is regulated by several factors, including ROS, inflammation, and interactions with neighboring cells such as cancer-associated fibroblasts and immune cells. Dysregulation of NRF2 activity can lead to an imbalance in the functioning of cellular functions, thus promoting tumor growth, metastasis, and resistance to chemotherapy and radiotherapy [12].

The tumoral changes in normal cells show eight main characteristics: unrestrained proliferation, avoided growth suppressor signaling, resistance to cell death, sustained replication, induction of angiogenesis, activation of invasion and metastasis, and altered cellular metabolism, including redox homeostasis and evading immune destruction. All of these characteristics are part of what they called Hallmarks of cancer (**Figure 2**) [13].

Although NRF2 is considered one of the main molecules that are part of the activation of the cellular antioxidant response, its role in carcinogenesis remains controversial. Nrf2 deficiency seems to arrest the cell cycle in G2/M phase, indicating that Nrf2 is required for cell cycle regulation [14]. In fact, the effects of NRF2 in tumor growth and progression have been established in multiple types of human cancers, showing that proliferation rates of cell lines vary according to their NRF2 status [15]. NRF2 can prevent carcinogenesis by facilitating the rapid enzymatic modification and elimination of chemical carcinogens, as well as by neutralizing ROS or repairing oxidative damage through the regulation of the expression of its target genes [16]. On the other hand, NRF2 overactivation can be associated with the initiation and spread of tumors, encouraging their growth, migration, and invasion as well as providing therapeutic and stress resistance [17].

NRF2 governs the expression of major enzymes involved in the maintenance of redox homeostasis, facilitating proliferative response [17]. NRF2 regulates several genes related to the cell cycle such as neurogenic locus notch homolog protein 1 (NOTCH1),

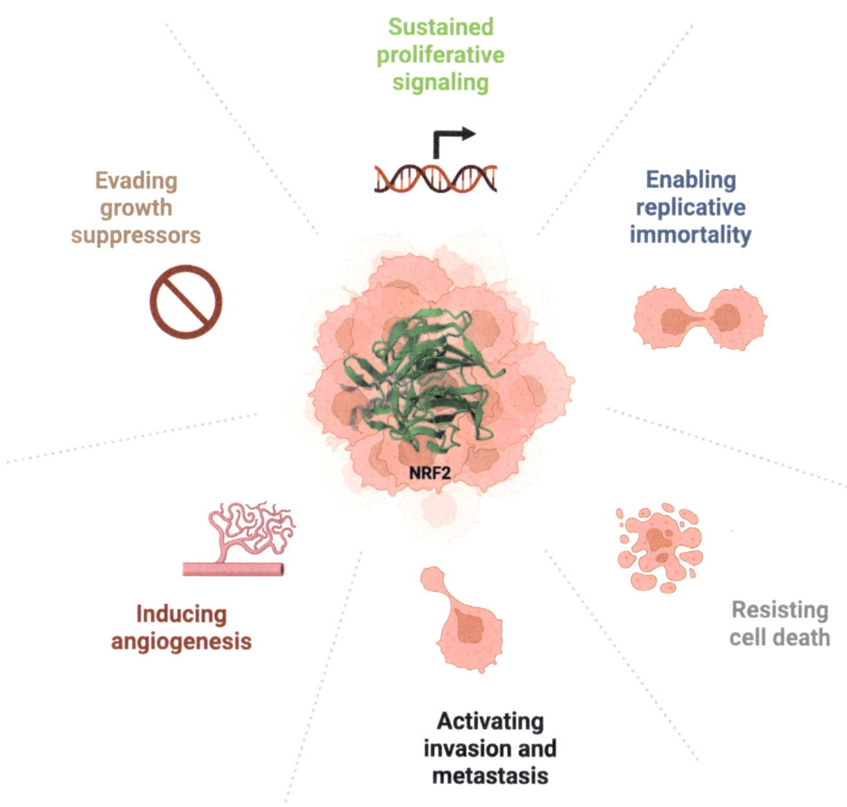

Figure 2.
NRF2 and hallmarks of cancer.

Nephronectin (NPNT), bone morphogenetic protein receptor, type IA (BMPR1A), insulin-like growth factor 1 (IFG1), integrin beta 2 (ITGB2), platelet-derived growth factor C (PDGFC), vascular endothelial growth factor C (VEGFC), and Jagged Canonical Notch Ligand 1 (JAG1) [18], most of them involved in cancer maintenance. So, activation of Nrf2 not only protects the cancer cells from antioxidants but also influences their surrounding microenvironment. Although cells that do not express NRF2 show slower proliferation [15], NRF2 activation in the tumor microenvironment was found to slow the progression of previously activated malignant tumors instead of promoting tumor growth directly. These findings challenge the conventional view of this molecule exclusively promoting tumor growth, reinforcing its dual role [19].

Redox signaling pathways in immune cells are complex mechanisms. The expression and signaling of NRF2 may control the activity of redox-sensitive immunoregulatory molecules that affect immune cell subsets' functional status [20]. NRF2 is an important regulator of anti-inflammatory gene expression, and its effect on immune responses can vary depending on the context. Activation of NRF2 has been demonstrated to decrease inflammation through lowering ROS levels as well as proinflammatory cytokine production, including IL-1β, IL-6, and TNF-α [21].

The participation of Nrf2 in the tumor microenvironment is also linked to several interactions between tumor-associated macrophages (TAMs) and tumor cells within the microenvironment, culminating in an immunosuppressive environment that promotes tumor growth and invasion [22]. The activation of NFR2 in cancer cells also affects macrophage polarization toward M2-like populations, which are linked to the development of tumors [23].

NRF2 is necessary for normal endothelial angiogenic processes, as it maintains a normal endothelium phenotype and vascular integrity [24]. NRF2 knockdown decreases the amounts of HIF-1α protein, a transcription factor stimulated by the tumor hypoxic microenvironment, which sets off a signaling cascade that stimulates the transcription of extracellular matrix (ECM) remodelers, cytokines, and growth factors to produce vasculature [25]. Therefore, Nrf2 acts to decrease the expression of vascular endothelial growth factor (VEGF), platelet-derived growth factor (PDGF), angiopoietin, and angiogenin, important molecules in the angiogenesis process [26]. Under stressful circumstances, NRF2 also regulates Bcl-2 expression, an antiapoptotic protein whose overexpression is usually associated with poor prognosis in many human cancers [27]. Then Nrf2 could also inhibit cellular death, resulting in cell survival, and influence drug resistance [28].

Numerous mechanisms affect the regulation of NRF2, and the way its activation determines the signaling and contribution to either cancer progression or prevention. These mechanisms include canonical activation, autophagy dysfunction, transcriptional and translational modifications, epigenetics, post-translational modifications, protein-protein interactions, and mutations [29]. The neoplastic cells modify the core metabolism by upregulating important metabolic pathways such as glycolysis, the pentose phosphate pathway (PPP), and glutaminolysis [30], reprogramming the cellular events, one of the important hallmarks of cancer [31]. It is a remarkable contribution to the rapid proliferation of cancer, at least by supporting biosynthetic needs.

All these mechanisms, including epigenetic modifications, have a significant impact on cancer by upregulating NRF2 and enabling drug resistance [32]. However, it is important to consider the role of the tumor microenvironment in cancer therapy to better understand how cancer cells interact with their surroundings, which can thus provide valuable insights for the development of more effective and precise therapeutic strategies [19].

3. NRF2 in female cancer

3.1 Breast cancer

Breast cancer is the most frequently diagnosed cancer and the leading cause of cancer death in females worldwide, with an estimated 2.3 million new cancer cases in 2020. The role of Nrf2 in different subtypes of breast cancer is still being investigated by various researchers. To date, few data have been published regarding the level of Nrf2 expression in different types of breast cancer. However, it is known that Nrf2 and Keap-1 pathways are more activated in breast cancer cases positive for estrogen (ER), progesterone (PR), and human epidermal growth factor receptor 2 (HER2) receptors compared to triple-negative breast cancer (TNBC) [33]. Although in healthy cells, Nrf2 may prevent cancer by suppressing DNA damage and carcinogenesis triggered by ROS, high NRF2 expression in breast cancer patients regulates cancer-related processes and is correlated with lower survival rates [34].

NRF2 silencing was related to decreased expression of molecules like small GTPase and RhoA protein, along with the reduction of breast cancer cell proliferation, chemotaxis, and cell migration. On the contrary, NRF2 activation affects the RhoA/ROCK signaling pathway, probably through its binding to the estrogen-related receptor alpha (ERR1) promoter, preventing RhoA degradation and promoting its pro-oncogenic signaling [35]. In ER-negative breast cancer, Nrf2 overexpression negatively regulates C-X-C Motif Chemokine Ligand 13 (CXCL13), thus suppressing cancer proliferation, while the increased level of CXCL13/CXCR5 in ER-positive breast cancer cells helps advance tumor invasion and metastasis. In TNBC, not only NRF2 mRNA levels but also Keap1 have been demonstrated to be regulated by several microRNAs, like miR-200, that lowers KEAP1 mRNA levels and increases the NRF2 expression [36].

Evidence indicates that NRF2 plays a crucial role in activating protective mechanisms against oxidative stress or toxins, yet sustained NRF2 signaling in cancer cells can promote oncogenic programs, leading to therapy resistance and poor outcomes [37]. However, due to the dual role of Nrf2 in breast cancer patients, a better understanding of NRF2-related events is necessary [36].

3.2 Cervical cancer

Cancers of the female reproductive system are an important cause of cancer morbidity and mortality worldwide. Cervical, endometrial, and ovarian cancers are relatively common and cause significant cancer morbidity and mortality worldwide, whereas vulvar, vaginal, fallopian tube cancers, and choriocarcinomas are very rare [38]. Cervical cancer is the fourth most frequently diagnosed malignant disease among women worldwide, with 604,000 new cases in 2020 [39], mainly caused by human papillomavirus (HPV) infection [40]. The integration of the viral genome with the human host is essential for cancer progress, and oxidative stress plays an important role in viral integration and persistent virus infection [41].

The cervical tissue is exposed to oxidative stress, which can lead to the development and increased neoplastic transformation of the tissue [42]. Perhaps NRF2 in cervical cancer maintains the redox balance in the cell by controlling gene transcription [43]. In fact, NRF2's antioxidant response is crucial for redox balance during HPV infection, preventing oxidative stress and DNA damage and maintaining genomic stability. Alterations in the Nrf2/Keap pathway during HPV-related cancer progression can lead to Nrf2 overexpression and activation, boosting antioxidant enzymes like nicotinamide adenine dinucleotide phosphate (NADPH) quinone oxidoreductase 1 (NQO1), heme oxygenase-1 (HO-1), glutathione S-transferase (GST), catalase, and superoxide dismutase (SODs). This reduces oxidative stress and chemotherapy toxicity, aiding therapy resistance [41].

Nrf2 has a role in promoting cell metastasis in cervical cancer once its activation influences various aspects involved in metastasis, such as cell mobility, invasion, immune evasion, and colonization of distant sites. Nrf2 enhanced Epithelial-mesenchymal transition (EMT) and anoikis resistance by promoting the expression of Snail1 and the metastasis of HeLa and SiHa cervical cancer cells [44]. NRF2 also showed a strong nuclear expression associated with reduced cytoplasmic Keap1 expression in cervical cancers, being related to a hypermethylation phenotype. The results also demonstrate that cell migration abilities were enhanced following NRF2 overexpression compared to the control group. Additionally, SiHa cells with reduced NRF2 expression show inhibited migration, while overexpression increases the invasive behavior. Then, inhibition of NRF2, in combination with antineoplastic agents, would be a promising therapeutic strategy to inhibit metastasis in cervical cancer [43, 44].

3.3 Endometrial and ovarian cancer

Worldwide, endometrial cancer stands as the predominant gynecologic malignancy in affluent nations, with more than 417,000 new cases of endometrial cancer and 314,000 new ovarian cancer cases in 2020. Although factors such as aging demographic and decline in benign hysterectomies have played a role in endometrial cancer rates, the primary driving force behind this is the increasing prevalence of obesity, a factor that may also contribute to ovarian cancer rate [45].

There are two subtypes of endometrial cancer, according to histological findings. Type I is classified as endometrioid, while type II consists of serous and clear cell carcinomas. Endometrial cancer that shows serous and clear cell histologies shows aggressive nature and high recurrence rates. In a study involving hysterectomy specimens and cell culture, researchers discovered heightened Nrf2 levels in endometrial serous carcinoma (ESC), in contrast to minimal or absent expression in complex hyperplasia and endometrial endometrioid carcinoma (EEC). Additionally, ESC-derived cells exhibited increased Nrf2 expression and greater resistance to conventional chemotherapeutic agents compared to EEC-derived cells. Further experiments revealed that silencing Nrf2 expression sensitized ESC cells to chemotherapy [46]. High levels of Nrf2 were identified in a significant portion of endometrioid carcinoma cases with phosphatase and tensin homolog (PTEN) loss [47], showing the intricate relation of Nrf2 and the proliferative pathways. Given NRF2's role as a transcription factor with established target genes, Beinse et al. examined the mRNA levels of three such targets (NQO1, GCLC, and AKR1C3), showing that serous endometrial cancer exhibits low NQO1 expression, potentially linked to a collaboration between NRF2 and TP53, driving their aggressiveness [48].

Regarding ovarian cancers, the majority are epithelial ovarian cancers, but different histological subtypes have been identified, including serous (the most common), endometrioid, clear cell, and mucinous cancers [49]. Nrf2 was also related to ovarian cancer once NQO1 and NRF2 were highly expressed compared to normal and precancerous lesions, showing a positive correlation across different types of lesions. NRF2 expression was observed in both the nucleus and cytoplasm of ovarian cells, with a significant increase observed as the ovarian carcinoma stage advances [50]. NRF2 is correlated to progesterone receptor (PR) expression across all histological subtypes, with a strong correlation between cytoplasmic NRF2 and the expression of progesterone receptor A (PRA) and progesterone receptor B (PRB), associated with longer overall survival [51].

4. NFR2 as target for cancer treatment

NRF2 possesses the remarkable ability to regulate the expression of numerous cytoprotective genes, effectively shielding cells and tissues from diverse threats, spanning environmental insults to metabolic dysregulation. Yet, when the NRF2 pathway deviates from its regular course, aberrant activation emerges as a formidable ally to cancer cells, bolstering their survival, proliferation, and resistance to treatment modalities [12]. Consequently, this transcription factor has garnered significant attention as a promising therapeutic target across various cancer types.

The NRF2 pathway in cancer cells can lead to therapy resistance through the upregulation of drug-metabolizing enzymes and efflux transporters. Consequently, targeting NRF2 has become a focus for therapeutic strategies. One approach involves modulating KEAP1 to disrupt NRF2 activity positively [52]. Another approach is inhibiting NRF2 directly. Screening has been valuable in identifying small molecule inhibitors of NRF2 activity, such as AEM1 and ML385, which have shown efficacy in preclinical models of non-small cell lung cancer (NSCLC) [53].

NRF2-induced anticancer resistance can be observed in many types of cancers for a variety of drugs, like cisplatin, doxorubicin, 5-fluorouracil, paclitaxel, and bleomycin [54]. Only few molecules were found to inhibit Nrf2, but pharmacological compounds could indirectly drive down the Nrf2 overactivation effects and sensitize the neoplastic cells to chemotherapy. Then, screening efforts have uncovered potential molecular targets for NRF2 inhibition, including Twist Family basic helix-loop-helix (BHLH) Transcription Factor 1 (TWIST1), E74 Like (E26 transformation-specific) ETS Transcription Factor 4 (ELF4), Serine/threonine-protein kinase Nek8 (NEK8), TGF-Beta Activated Kinase 1 (MAP3K7) Binding Protein 1 (TAB1), and Dual Specificity Phosphatase 4 (DUSP4), expanding the possibilities for therapeutic intervention [53].

Pan-cancer therapies are innovative treatments designed to target shared molecular pathways or vulnerabilities across different types of cancer. By focusing on common genetic mutations or dysregulated pathways, these therapies offer a promising approach to treating multiple cancers. Examples include immunotherapy, which harnesses the immune system to fight cancer and targeted therapy, which inhibits specific molecular targets. Advancements in precision medicine have further enhanced the development of tailored therapies based on individual tumor profiles. A pan-cancer analysis shows immunoevasive characteristics in NRF2 hyperactive squamous malignancies, unveiling an intriguing phenotype in squamous malignancies of a variety of cancers, including cervical cancer, characterized by high NRF2 activity associated with diminished interferon-gamma (IFNγ), HLA-I expression, and reduced TAMs. Moreover, the study identified upregulated immunomodulatory genes such as Nicotinamide phosphoribosyltransferase (NAMPT), Wnt Family Member 5A (WNT5A), Secreted Phosphoprotein 1 (SPP1), solute carrier family 7 member 11 (SLC7A11), solute carrier family 2 member 1 (SLC2A1), and Programmed death-ligand 1 (PD-L1) in these NRF2 hyperactive diseases, suggesting novel potential targets for NRF2 modulation to alter the tumor immune microenvironment [55].

5. Conclusion

NRF2 has a dual role in carcinogenesis, acting as protective in normal cells, but worsening the status of the neoplastic cells. NRF2 also plays a role in chemoresistance by reducing drug deposits inside cancer cells and preventing the cells from dying. Once one of the major issues for effective treatment of many tumors is chemotherapy resistance, NRF2 inhibition may be a useful strategy to improve the health support for cancer patients. However, it is important to highlight the need for understanding the right Nrf2 level range to be regulated in order to avoid strong side effects caused by any of the proposed targeted therapies.

Author details

Danyelly Bruneska Gondim Martins[1,2]*, Ananda Cristina de Aguiar[1], Francielle Maria de Araújo Barbosa[1] and Glauber Moreira Leitão[3]

1 Keizo Asami Institute (iLIKA), Federal University of Pernambuco, Recife, Pernambuco, Brazil

2 Department of Biochemistry, Federal University of Pernambuco, Recife, Pernambuco, Brazil

3 Academic Area of Clinical Medicine, Center for Medical Sciences, School of Medicine of Recife, Federal University of Pernambuco – UFPE, Recife, Pernambuco, Brazil

*Address all correspondence to: bruneska@prospecmol.org

IntechOpen

© 2024 The Author(s). Licensee IntechOpen. This chapter is distributed under the terms of the Creative Commons Attribution License (http://creativecommons.org/licenses/by/3.0), which permits unrestricted use, distribution, and reproduction in any medium, provided the original work is properly cited.

References

[1] Sies H, Berndt C, Jones DP. Oxidative stress. Annual Review of Biochemistry. 2017;**86**:715-748. DOI: 10.1146/annurev-biochem-061516-045037

[2] Canning P, Sorrell FJ, Bullock AN. Structural basis of Keap1 interactions with Nrf2. Free Radical Biology & Medicine. 2015;**88**(Pt B):101-107. DOI: 10.1016/j.freeradbiomed.2015.05.034

[3] Taguchi K, Yamamoto M. The KEAP1-NRF2 system in cancer. Frontiers in Oncology. 2017;**4**(7):85. DOI: 10.3389/fonc.2017.00085

[4] Tong KI, Katoh Y, Kusunoki H, Itoh K, Tanaka T, Yamamoto M. Keap1 recruits Neh2 through binding to ETGE and DLG motifs: Characterization of the two-site molecular recognition model. Molecular and Cellular Biology. 2006;**26**(8):2887-2900

[5] Chowdhry S, Zhang Y, McMahon M, Sutherland C, Cuadrado A, Hayes JD. Nrf2 is controlled by two distinct β-TrCP recognition motifs in its Neh6 domain, one of which can be modulated by GSK-3 activity. Oncogene. 2013;**32**(32):3765-3781. DOI: 10.1038/onc.2012.388

[6] Wang H, Liu K, Geng M, Gao P, Wu X, Hai Y, et al. RXRα inhibits the NRF2-ARE signaling pathway through a direct interaction with the Neh7 domain of NRF2. Cancer Research. 2013;**73**(10):3097-3108. DOI: 10.1158/0008-5472.CAN-12-3386

[7] Itoh K, Chiba T, Takahashi S, Ishii T, Igarashi K, Katoh Y, et al. An Nrf2/small Maf heterodimer mediates the induction of phase II detoxifying enzyme genes through antioxidant response elements. Biochemical and Biophysical Research Communications. 1997;**236**(2):313-322. DOI: 10.1006/bbrc.1997.6943

[8] Ulasov AV, Rosenkranz AA, Georgiev GP, Sobolev AS. Nrf2/Keap1/ARE signaling: Towards specific regulation. Life Sciences. 2022;**291**:120111. DOI: 10.1016/j.lfs.2021.120111

[9] Jiang T, Harder B, Rojo de la Vega M, Wong PK, Chapman E, Zhang DD. p62 links autophagy and Nrf2 signaling. Free Radical Biology & Medicine. 2015;**88**(Pt B):199-204. DOI: 10.1016/j.freeradbiomed.2015.06.014

[10] Kitamura H, Motohashi H. NRF2 addiction in cancer cells. Cancer Science. 2018;**109**(4):900-911. DOI: 10.1111/cas.13537

[11] Basak P, Sadhukhan P, Sarkar P, Sil PC. Perspectives of the Nrf-2 signaling pathway in cancer progression and therapy. Toxicology Reports. 2017;**13**(4):306-318. DOI: 10.1016/j.toxrep.2017.06.002

[12] Rojo de la Vega M, Chapman E, Zhang DD. NRF2 and the hallmarks of cancer. Cancer Cell. 2018;**34**(1):21-43. DOI: 10.1016/j.ccell.2018.03.022

[13] Hanahan D, Weinberg RA. Hallmarks of cancer: The next generation. Cell. 2011;**144**(5):646-674. DOI: 10.1016/j.cell.2011.02.013

[14] Reddy NM, Kleeberger SR, Bream JH, Fallon PG, Kensler TW, Yamamoto M, et al. Genetic disruption of the Nrf2 compromises cell-cycle progression by impairing GSH-induced redox signaling. Oncogene. 2008;**27**(44):5821-5832. DOI: 10.1038/onc.2008.188

[15] Zhang M, Zhang C, Zhang L, Yang Q, Zhou S, Wen Q, et al. Nrf2 is a potential prognostic marker and

promotes proliferation and invasion in human hepatocellular carcinoma. BMC Cancer. 2015;**15**:531. DOI: 10.1186/s12885-015-1541-1

[16] Harder B, Jiang T, Wu T, Tao S, Rojo de la Vega M, Tian W, et al. Molecular mechanisms of Nrf2 regulation and how these influence chemical modulation for disease intervention. Biochemical Society Transactions. 2015;**43**(4):680-686. DOI: 10.1042/BST20150020

[17] Zimta AA, Cenariu D, Irimie A, Magdo L, Nabavi SM, Atanasov AG, et al. The role of Nrf2 activity in cancer development and progression. Cancers (Basel). 2019;**11**(11):1755. DOI: 10.3390/cancers11111755

[18] Wakabayashi N, Shin S, Slocum SL, Agoston ES, Wakabayashi J, Kwak MK, et al. Regulation of notch1 signaling by nrf2: Implications for tissue regeneration. Science Signaling. 2010;**3**(130):ra52. DOI: 10.1126/scisignal.2000762

[19] Hayashi M, Kuga A, Suzuki M, Panda H, Kitamura H, Motohashi H, et al. Microenvironmental activation of Nrf2 restricts the progression of Nrf2-activated malignant tumors. Cancer Research. 2020;**80**(16):3331-3344. DOI: 10.1158/0008-5472.CAN-19-2888

[20] Pant T, Uche N, Juric M, Zielonka J, Bai X. Regulation of immunomodulatory networks by Nrf2-activation in immune cells: Redox control and therapeutic potential in inflammatory diseases. Redox Biology. 2024;**70**:103077. DOI: 10.1016/j.redox.2024.103077

[21] Saha S, Buttari B, Panieri E, Profumo E, Saso L. An overview of Nrf2 signaling pathway and its role in inflammation. Molecules. 2020;**25**(22):5474. DOI: 10.3390/molecules25225474

[22] Kobayashi EH, Suzuki T, Funayama R, Nagashima T, Hayashi M, Sekine H, et al. Nrf2suppressesmacrophageinflammatory response by blocking proinflammatory cytokine transcription. Nature Communications. 2016;**7**:11624. DOI: 10.1038/ncomms11624

[23] Feng R, Morine Y, Ikemoto T, Imura S, Iwahashi S, Saito Y, et al. Nrf2 activation drive macrophages polarization and cancer cell epithelial-mesenchymal transition during interaction. Cell Communication and Signaling: CCS. 2018;**16**(1):54. DOI: 10.1186/s12964-018-0262-x

[24] Guo Z, Mo Z. Keap1-Nrf2 signaling pathway in angiogenesis and vascular diseases. Journal of Tissue Engineering and Regenerative Medicine. 2020;**14**(6):869-883. DOI: 10.1002/term.3053

[25] Teleanu RI, Chircov C, Grumezescu AM, Teleanu DM. Tumor angiogenesis and anti-angiogenic strategies for cancer treatment. Journal of Clinical Medicine. 2019;**9**(1):84. DOI: 10.3390/jcm9010084

[26] Hamada S, Matsumoto R, Masamune A. HIF-1 and NRF2; key molecules for malignant phenotypes of pancreatic cancer. Cancers (Basel). 2022;**14**(2):411. DOI: 10.3390/cancers14020411

[27] Qian S, Wei Z, Yang W, Huang J, Yang Y, Wang J. The role of BCL-2 family proteins in regulating apoptosis and cancer therapy. Frontiers in Oncology. 2022;**12**:985363. DOI: 10.3389/fonc.2022.985363

[28] Niture SK, Jaiswal AK. Nrf2-induced antiapoptotic Bcl-xL protein enhances cell survival and drug resistance. Free Radical Biology & Medicine. 2013;**57**:119-131. DOI: 10.1016/j.freeradbiomed.2012.12.014

[29] Li R, Jia Z, Zhu H. Regulation of Nrf2 Signaling. Reactive Oxygen Species (Apex). 2019;**8**(24):312-322

[30] Schiliro C, Firestein BL. Mechanisms of metabolic reprogramming in cancer cells supporting enhanced growth and proliferation. Cells. 2021;**10**(5):1056. DOI: 10.3390/cells10051056

[31] Pavlova NN, Zhu J, Thompson CB. The hallmarks of cancer metabolism: Still emerging. Cell Metabolism. 2022;**34**(3):355-377. DOI: 10.1016/j.cmet.2022.01.007

[32] Bouyahya A, Mechchate H, Oumeslakht L, Zeouk I, Aboulaghras S, Balahbib A, et al. The role of epigenetic modifications in human cancers and the use of natural compounds as Epidrugs: Mechanistic pathways and pharmacodynamic actions. Biomolecules. 2022;**12**(3):367. DOI: 10.3390/biom12030367

[33] Karihtala P, Kauppila S, Soini Y, Arja-Jukkola-Vuorinen. Oxidative stress and counteracting mechanisms in hormone receptor positive, triple-negative and basal-like breast carcinomas. BMC Cancer. 2011;**11**:262. DOI: 10.1186/1471-2407-11-262

[34] Tascioglu Aliyev A, Panieri E, Stepanić V, Gurer-Orhan H, Saso L. Involvement of NRF2 in breast cancer and possible therapeutical role of polyphenols and melatonin. Molecules. 2021;**26**(7):1853. DOI: 10.3390/molecules26071853

[35] Zhang C, Wang HJ, Bao QC, Wang L, Guo TK, Chen WL, et al. NRF2 promotes breast cancer cell proliferation and metastasis by increasing RhoA/ROCK pathway signal transduction. Oncotarget. 2016;**7**(45):73593-73606. DOI: 10.18632/oncotarget.12435

[36] Kumar H, Kumar RM, Bhattacharjee D, Somanna P, Jain V. Role of Nrf2 signaling cascade in breast cancer: Strategies and treatment. Frontiers in Pharmacology. 2022;**13**:720076. DOI: 10.3389/fphar.2022.720076

[37] Almeida M, Soares M, Ramalhinho AC, Moutinho JF, Breitenfeld L, Pereira L. The prognostic value of NRF2 in breast cancer patients: A systematic review with meta-analysis. Breast Cancer Research and Treatment. 2020;**179**(3):523-532. DOI: 10.1007/s10549-019-05494-4

[38] Weiderpass E, Labrèche F. Malignant tumors of the female reproductive system. Safety and Health at Work. 2012;**3**(3):166-180. DOI: 10.5491/SHAW.2012.3.3.166

[39] Olawaiye AB, Baker TP, Washington MK, Mutch DG. The new (version 9) American joint committee on cancer tumor, node, metastasis staging for cervical cancer. CA: A Cancer Journal for Clinicians. 2021;**71**(4):287-298. DOI: 10.3322/caac.21663

[40] zur Hausen H. Papillomaviruses and cancer: From basic studies to clinical application. Nature Reviews. Cancer. 2002;**2**(5):342-350. DOI: 10.1038/nrc798

[41] Cruz-Gregorio A, Aranda-Rivera AK, Pedraza-Chaverri J. Nuclear factor erythroid 2-related factor 2 in human papillomavirus-related cancers. Reviews in Medical Virology. 2022;**32**(3):e2308. DOI: 10.1002/rmv.2308

[42] De Marco F. Oxidative stress and HPV carcinogenesis. Viruses. 2013;**5**(2):708-731. DOI: 10.3390/v5020708

[43] Ma JQ, Tuersun H, Jiao SJ, Zheng JH, Xiao JB, Hasim A. Functional role of NRF2 in cervical carcinogenesis. PLoS One. 2015;**10**(8):e0133876. DOI: 10.1371/journal.pone.0133876

[44] Zhang M, Hong X, Ma N, Wei Z, Ci X, Zhang S. The promoting effect and

mechanism of Nrf2 on cell metastasis in cervical cancer. Journal of Translational Medicine. 2023;**21**(1):433. DOI: 10.1186/s12967-023-04287-0

[45] Sung H, Ferlay J, Siegel RL, Laversanne M, Soerjomataram I, Jemal A, et al. Global cancer statistics 2020: GLOBOCAN estimates of incidence and mortality worldwide for 36 cancers in 185 countries. CA: a Cancer Journal for Clinicians. 2021;**71**(3):209-249. DOI: 10.3322/caac.21660

[46] Jiang T, Chen N, Zhao F, Wang XJ, Kong B, Zheng W, et al. High levels of Nrf2 determine chemoresistance in type II endometrial cancer. Cancer Research. 2010;**70**(13):5486-5496. DOI: 10.1158/0008-5472.CAN-10-0713

[47] Rojo AI, Rada P, Mendiola M, Ortega-Molina A, Wojdyla K, Rogowska-Wrzesinska A, et al. The PTEN/NRF2 axis promotes human carcinogenesis. Antioxidants & Redox Signaling. 2014;**21**(18):2498-2514. DOI: 10.1089/ars.2014.5843

[48] Beinse G, Just PA, Rance B, Izac B, Letourneur F, Saidu NEB, et al. The NRF2 transcriptional target NQO1 has low mRNA levels in TP53-mutated endometrial carcinomas. PLoS One. 2019;**14**(3):e0214416. DOI: 10.1371/journal.pone.0214416

[49] Rojas V, Hirshfield KM, Ganesan S, Rodriguez-Rodriguez L. Molecular characterization of epithelial ovarian cancer: Implications for diagnosis and treatment. International Journal of Molecular Sciences. 2016;**17**(12):2113. DOI: 10.3390/ijms17122113

[50] Osman NATAG, Abd El-Maqsoud NMR, El Gelany SAA. Correlation of NQO1 and Nrf2 in female genital tract cancer and their precancerous lesions (cervix, endometrium and ovary). World Journal of Oncology. 2015;**6**(3):364-374. DOI: 10.14740/wjon931w

[51] Czogalla B, Kahaly M, Mayr D, Schmoeckel E, Niesler B, Hester A, et al. Correlation of NRF2 and progesterone receptor and its effects on ovarian cancer biology. Cancer Management and Research. 2019;**14**(11):7673-7684. DOI: 10.2147/CMAR.S210004

[52] Xu Y, Fang F, Miriyala S, Crooks PA, Oberley TD, Chaiswing L, et al. KEAP1 is a redox sensitive target that arbitrates the opposing radiosensitive effects of parthenolide in normal and cancer cells. Cancer Research. 2013;**73**(14):4406-4417. DOI: 10.1158/0008-5472.CAN-12-4297

[53] Panieri E, Saso L. Potential applications of NRF2 inhibitors in cancer therapy. Oxidative Medicine and Cellular Longevity. 2019;**11**(2019):8592348. DOI: 10.1155/2019/8592348

[54] Choi B, Kwak M-K. Shadows of NRF2 in cancer: Resistance to chemotherapy. Current Opinion in Toxicology. 2016;**1**:20-28. DOI: 10.1016/j.cotox.2016.08.003

[55] Härkönen J, Pölönen P, Deen AJ, Selvarajan I, Teppo HR, Dimova EY, et al. A pan-cancer analysis shows immunoevasive characteristics in NRF2 hyperactive squamous malignancies. Redox Biology. 2023;**61**:102644. DOI: 10.1016/j.redox.2023.102644

Chapter 5

Nrf2 as a Therapy Target for Th17-Dependent Autoimmune Disease

Hongmin Wu and Xiancai Zhong

Abstract

Th17 cells are a subset of IL-17-expressing $CD4^+$ T helper cells and play a predominant role in the pathogenesis of autoimmune diseases such as multiple sclerosis and psoriasis. Th17 cells sustain their activation and effector functions primarily through a metabolic profile characterized by high glycolytic and oxidative metabolism. Both glycolysis and OXPHOs can affect cellular redox status, and vice versa. Nrf2, a master regulator of redox homeostasis, plays a pivotal role in oxidative stress regulation and influences immune cell function. This chapter summarizes the recent advances in the understanding of redox regulation in Th17 cells and explores the therapeutic potential of targeting Nrf2 in Th17-dependent autoimmune diseases. Overall, targeting Nrf2 holds considerable promise as a novel therapeutic paradigm for Th17-dependent autoimmune diseases, offering new avenues for precision medicine and improved disease outcomes.

Keywords: autoimmune disease, Th17, Nrf2, metabolic regulation, glycolysis, OXPHOs

1. Introduction

Autoimmune diseases, characterized by immune system dysfunction and self-directed attacks on healthy tissues, represent a significant challenge in clinical medicine. Among the various immune cell subsets implicated in autoimmunity, Th17 cells have emerged as key players due to their potent pro-inflammatory properties [1]. Th17 cells represent a distinct subset of $CD4^+$ T lymphocytes characterized by their production of interleukin-17 (IL-17) cytokines. Originally identified for their role in host defense against extracellular pathogens, mounting evidence implicates dysregulated Th17 cell responses in the pathogenesis of autoimmune disorders, including multiple sclerosis (MS), rheumatoid arthritis (RA), and psoriasis [1]. Their pro-inflammatory properties contribute to tissue inflammation, damage, and chronicity, highlighting the importance of understanding the regulation and function of Th17 cells in health and disease [2]. Thus, there is an imperative clinical demand for therapies capable of efficiently managing Th17-mediated autoimmunity.

Studies have shown that Th17 cells utilize both glycolysis and oxidative phosphorylation (OXPHOs) as energy sources [3, 4]. During the glycolysis and OXPHOs, reactive oxygen species (ROS) are produced as by-products [5]. Given the high metabolic activity of differentiated Th17 cells, there may be more production of ROS during Th17 differentiation. This heightened ROS production can perturb cellular redox balance and potentially lead to oxidative stress within the cellular milieu [6]. ROS-mediated oxidative damage to biomolecules such as lipids, proteins, and DNA can exacerbate tissue inflammation and further contribute to Th17-mediated immunity and the development of autoimmune diseases [7].

Nuclear factor erythroid 2-related factor 2 (Nrf2) is a transcription factor renowned for its pivotal role in cellular defense mechanisms. Nrf2 orchestrates the expression of a multitude of genes involved in antioxidant response, detoxification, and cellular metabolism, thereby safeguarding cells from oxidative damage and maintaining redox homeostasis [8]. Emerging evidence suggests that Nrf2 exerts profound effects on immune cell function [7]. Various immune cell subsets, including macrophages, dendritic cells, T cells, and natural killer cells, have been shown to express Nrf2 and exhibit altered phenotypes upon its modulation. This modulation extends beyond traditional antioxidant and detoxification pathways, implicating Nrf2 in the regulation of immune cell development, activation, differentiation, and effector functions [9].

Unraveling the intricate interplay between Th17 cell differentiation, metabolic pathways, and cellular redox status has revealed promising therapeutic avenues. By modulating Nrf2 activity, either pharmacologically or through dietary interventions, it becomes feasible to regulate Th17 cell differentiation and function, thereby mitigating autoimmune pathology. This study will discuss evidence of Nrf2-mediated modulation of Th17 cells and underscore recent studies supporting the efficacy of Nrf2 activation as a therapeutic approach. Furthermore, it provides insights into Nrf2 as a potential therapy target for Th17-dependent autoimmune disease, offering a perspective on the development of novel treatment approaches in this context.

2. Metabolic regulation of Th17 cells

Th17 cells represent a distinct subset of $CD4^+$ T lymphocytes characterized by their production of IL-17 cytokines. Th17 cells have both non-pathogenic and pathogenic features. The non-pathogenic aspect helps to protect the host against external bacterial and fungal pathogens [10], while the pathogenic facet is involved in driving the progression of autoimmune diseases such as MS, RA, and psoriasis [1]. Compared to the quiescent and metabolically inactive naïve $CD4^+$ T cells, Th17 cells are metabolically active [4]. Th17 cells exhibit distinct metabolic profiles, with metabolic reprogramming essential for their differentiation, activation, and effector functions. The metabolic demands of Th17 cells vary depending on the context of activation. Homeostatic Th17 cells induced by commensal bacteria, such as segmented filamentous bacteria, primarily rely on OXPHOs [11]. In contrast, pro-inflammatory Th17 cells elicited in response to pathogens like *Citrobacter rodentium* exhibit high glycolytic activity [11]. Metabolic intermediates and signaling pathways not only fuel Th17 cell differentiation but also modulate cytokine production such as IL-17, IL-21, IL-22, and GM-CSF, thereby influencing autoimmune disease pathogenesis. Targeting metabolic pathways in Th17 cells offers promising therapeutic strategies to modulate autoimmune inflammation and restore immune balance. Although other metabolic processes such as lipid metabolism and amino acid metabolism are also involved in Th17 differentiation and serve as therapeutic targets for Th17-mediated autoimmune diseases [4], the current study mainly focuses on glycolysis and OXPHOs metabolic regulation in Th17 cells, summarizing their potential therapeutic strategies for Th17-related autoimmune diseases.

2.1 Glycolysis in Th17 differentiation

Glycolysis serves as the primary energy source in differentiated Th17 cells. During cell proliferation and differentiation, large amounts of glucose are transported into the Th17 cell to produce energy for various physiological processes. High glucose levels have been demonstrated to promote Th17 differentiation, exacerbating autoimmunity in mouse models [12]. The significance of glycolysis in Th17 cells has been underscored by numerous studies revealing dysregulated effector functions upon inhibition of glycolytic enzymes' activities or intermediate products. Genetic deficiency or pharmacological inhibition of key glycolytic molecules, such as hypoxia-inducible factor 1α (Hif1α), glucose transporter (GLUT), pyruvate kinase isoform 2 (PKM2), and pyruvate dehydrogenase kinase 1 (PDHK1), leads to significant impairments in Th17 cell differentiation and pro-inflammatory function [13–17].

HIF1α plays a crucial role in directing the differentiation of Th17 cells by directly activating the transcription of the Th17 master regulator RAR-related orphan receptor gamma (RORγt). Mice with T cells deficient in HIF1α demonstrate resistance to the induction of Th17-dependent experimental autoimmune encephalitis (EAE), a mouse model of autoimmune disease characterized by decreased Th17 cells and increased Treg cells [13]. In addition, the increased glycolytic rates in Th17 cells coincide with elevated expression of GLUT1, and increased GLUT1 levels are associated with elevated release of IL-17 and IFNγ [14]. Another glucose transporter protein, GLUT3, is also reported to be essential for both the differentiation and effector function of Th17 cells, as evidenced by the protection of GLUT3 knockout mice against EAE [15]. Furthermore, inhibition of PKM2 by vitamin B5 disrupts PKM2 phosphorylation and nuclear translocation, resulting in the inhibition of glycolysis and impaired Th17 cell differentiation. This inhibition leads to protection against EAE [16]. Moreover, PDHK1, a kinase that inhibits pyruvate dehydrogenase to suppress pyruvate oxidation and instead promotes pyruvate conversion to lactate, is essential for Th17 effector function. Blocking of PDHK1 with dichloroacetate suppresses glycolysis, Th17 differentiation, and attenuates EAE development [17].

Together, these findings highlight the critical role of glycolysis in regulation of Th17 cells and suggest that targeting glycolysis could offer therapeutic promise for alleviating Th17-based immune disorders.

2.2 OXPHOs in Th17 cell differentiation

Glycolysis and OXPHOs are two primary metabolic pathways responsible for generating energy within cells. Glycolysis fuels the tricarboxylic acid cycle (TCA) cycle and supports OXHPOs by generating pyruvate. The pyruvate generated from glycolysis is imported into the mitochondria and converted to acetyl-CoA. Acetyl-CoA then enters the TCA cycle to produce NADH and FADH2, which are utilized for generating ATP via OXPHOs. Conversely, OXPHOs can affect glycolysis by influencing metabolic intermediates (e.g., citrate) and cellular signaling pathways (e.g., AMPK signaling) [18, 19]. Thus, the two pathways are interconnected, working in concert to maintain cellular energy balance.

While glycolysis was thought to dominate Th17 cell metabolism, research has also highlighted the importance of OXPHOs in sustaining Th17 cell responses [20]. Studies have indicated that Th17 cells rely on OXPHOs for their effector responses [21]. Consistent with this observation, proliferating Th17 cells have been shown to rely on both electron transport chain (ETC) complex I and complex II activation for IL-17 production [22]. Additionally, estrogen receptor alpha signaling has been implicated in enhancing pathogenic Th17 cell differentiation by promoting ETC complex IV assembly and preserving OXPHOs [23]. In support of these studies, inhibiting OXPHOs has been found to effectively alleviate symptoms of colitis and psoriasis in murine models [21]. Furthermore, treatment with oligomycin, an ETC complex V inhibitor, impaired glycolytic functions and the TCA cycle in Th17 cells, resulting in reduced expression of IL-17, IL-23, and Tgfβ3. Importantly, oligomycin-treated Th17 cells exhibited reduced EAE severity upon adoptive transfer *in vivo* [20]. Interestingly, oligomycin treatment also promoted the generation of Treg cells via upregulating expression of IL-10 receptor [20].

Taken together, mitochondrial functions and OXPHOs contribute to Th17 cell generation. However, it is noteworthy that, compared to studies focusing on glycolysis, there is a relatively limited amount of research investigating the role of OXPHOs in Th17 cell differentiation and function. Thus, more studies are required to fully elucidate the impact of OXPHOs on Th17 cells and their contribution to immune responses.

3. Interplay between ROS, metabolic flux, and Th17-mediated autoimmunity

ROS are highly unstable, oxygen-containing molecules. ROS encompasses superoxide anion radical, hydroxyl radical, and hydroperoxyl radical [7]. ROS are produced from various sources, and one of the most significant sources is OXPHOs. Upon TCR activation, calcium ions exit from endoplasmic reticulum and influx into mitochondria to promote TCA cycle activity and increase mitochondrial function [5]. Increased TCA cycle leads to increased production of redox coenzymes NADH and FADH2. NADH and FADH2 are the main drivers of ETC by delivering electrons to the ETC in mitochondria. As electrons move through mitochondrial complexes, some may inadvertently leak, leading to the generation of superoxide radicals. In cells with high metabolic states, such as proliferating T cells, more superoxide radicals can be produced, culminating in uncontrolled oxidative stress [6]. Such uncontrolled oxidative stress has been associated with the development of various human diseases, including autoimmune disorders [7].

The association between a high glycolysis flux and increased ROS levels has been established [24]. This association arises from the fact that an increase in glycolytic flux typically leads to a rise in TCA cycle flux. Consequently, there is an accumulation of substrates (e.g., NADH and FADH2) for OXPHOs, potentially stimulating superoxide production [25]. Consistently, in T cells, ADP-dependent glucokinase has been identified as a mediator connecting high glycolysis with production of ROS, elucidating the critical role of glycolytic enzymes in modulating ROS levels [26]. Inhibition of mitochondrial pyruvate uptake restored mitochondrial ROS production to normal levels, thereby blocking the hyperglycemic damage [27]. Moreover, under hyperglycemic conditions, ROS production is induced through the process of mitochondrial fission [28]. These findings collectively demonstrate that increased glycolysis indirectly contributes to ROS and mitochondrial dysfunction. Conversely, ROS can impact glycolysis, forming a feedback loop by stabilizing HIF1α and promoting cell proliferation [29]. Stabilized HIF1α further facilitates a metabolic shift favoring anaerobic glycolysis and reducing mitochondrial respiration by upregulating GLUTs and glycolytic enzymes [30]. Thus, there is an intricate interplay between glycolysis and ROS regulation.

Studies have explored the role of ROS in the differentiation and pathogenicity of Th17 cells. Specifically, deficiency in NADPH oxidase, which maintains low mitochondrial ROS levels, has been found to enhance Th17 differentiation and result in increased susceptibility to EAE [31]. Additionally, knockout of IEX-1 gene, which suppresses mitochondrial ROS production, leads to enhanced Th17 differentiation and exacerbation of arthritis symptoms in a mouse model [32]. Moreover, high glucose concentrations have been demonstrated to drive Th17 differentiation both *in vitro* and *in vivo* through enhancing mitochondrial ROS production [12]. Notably, N-acetylcysteine, a specific inhibitor of mitochondrial ROS production, has been reported to significantly inhibit Th17 differentiation [33]. These findings collectively highlight the significant impact of dysregulated redox balance on Th17 differentiation and pathogenicity, underscoring the necessity of proper regulation of ROS to control the development of Th17-related autoimmune diseases. Maintaining the balance between ROS and antioxidant proteins within the cell is critical for preserving the integrity of Th17 cell-mediated immunity. As Nrf2-mediated signaling is a vital antioxidant system for maintaining intracellular redox balance, targeting Nrf2 emerges as a promising therapeutic strategy for Th17-dependent autoimmune diseases.

4. Insights into Nrf2-targeted therapy for Th17-dependent autoimmune diseases

Nrf2 is a stress-sensitive transcription factor that serves as a critical regulator in maintaining cellular redox hemostasis. Upon oxidative stress, Nrf2 translocates from the cytoplasm to the nucleus, where it becomes activated. Once activated, Nrf2 binds to antioxidant response elements (ARE) located in the promoters of target genes, initiating the transcription of ARE-containing cytoprotective antioxidant genes [8]. These antioxidant response genes, downstream of Nrf2, play a pivotal role in maintaining cellular redox balance within a physiological range. This delicate balance ensures proper metabolic processes and is essential for cellular function and viability [7].

Reports have elucidated the impact of Nrf2 on Th17 differentiation and pathogenicity. Deficiency in Nrf2 has been associated with augmented Th17 differentiation and the production of relevant cytokines (e.g., IL-17 and IL-23) both in *in vitro* experiments and in a murine model of lupus nephritis [34]. In addition, Nrf2-dependent antioxidant gene expression is suppressed in *in vitro* differentiated Th17 cells, and Nrf2 activation inhibits IL-17a production at both transcript and protein levels in peripheral blood mononuclear cells derived from MS patients. Mechanistically, Nrf2 binds to ARE repressor of *Rorc* promoter and inhibits *Rorc*-dependent IL-17 transactivation [35]. Conversely, T cell-specific overexpression of Nrf2 has been shown to promote the development of T regulatory cells [36]. These findings indicate the pivotal role of Nrf2 in modulating immune responses, as it not only prevents the differentiation of pro-inflammatory Th17 cells but also promotes the shift toward an anti-inflammatory T regulatory cell phenotype. Given its regulatory effects on Th17 differentiation and pathogenicity, Nrf2 emerges as a promising therapeutic target for Th17-mediated autoimmune disorders such as MS, psoriasis, and RA.

EAE is a widely used mouse model for MS research. It is induced by the activation of Th17 and Th1 cells, which penetrate the central nervous system, resulting in the degradation of myelin and death of oligodendrocytes [37]. Nrf2-deficient mice exhibit exacerbated EAE outcomes, marked by increased production of pro-inflammatory cytokines in the brain and spinal cord, and enhanced T cell infiltration in the central nervous system [38, 39]. Dimethyl fumarate (DMF) and its metabolite, monomethyl fumarate, demonstrate therapeutic potential by stabilizing Nrf2 and enhancing its transcriptional activity, effectively reducing the severity of EAE [40, 41]. Beyond DMF, other Nrf2 activators, like trifluoroethyl amide derivate of 2-cyano-3,12-dioxooleana-1,9(11)-dien-28-oic acid (CDDO-TFEA), exhibit promise in EAE management. CDDO-TFEA administration correlates with diminished levels of circulating pro-inflammatory cytokines and increased expression of Nrf2 and its target genes in various immune compartments, including brain, spinal cord, splenocytes, and CNS-infiltrating immune cells [42]. Moreover, compounds such as D3T [43] and TFM-735 [44], both of which activate Nrf2, have demonstrated protective effects during EAE with notable immunomodulatory functions.

RA is characterized by dysregulated Th17-driven inflammation. Studies have demonstrated the impact of Nrf2 on RA pathogenesis. Nrf2 deficiency increased leukocyte infiltration and accelerated the incidence of arthritis in mice [45]. Moreover, Nrf2-deficient mice exhibit more inflammation and oxidative damage in antibody-induced arthritis, a mouse model of RA [46]. Additionally, DMF treatment has been shown to attenuate osteoclastogenesis and bone destruction via induction of Nrf2-mediated transcription of antioxidant genes in a mouse model of LPS-induced arthritis [47]. Furthermore, treatment with S-propargyl-cysteine has demonstrated efficacy in reducing RA-associated inflammation by modulating the Nrf2-ARE signaling pathway [48]. These findings collectively suggest that targeting Nrf2 may provide therapeutic strategy for RA patients.

Th17 is believed to be the key contributor to the inflammatory processes underlying psoriasis, a chronic inflammatory skin disorder. Studies using Nrf2-deficient mice

Disease	Model	Effect of Nrf2	Reference
Mouse lupus nephritis	Nrf2 deficiency	increased Th17 differentiation, increased lupus nephritis	[34]
Mouse EAE	Nrf2 deficiency	exacerbated EAE	[38, 39]
Mouse RA	Nrf2 deficiency	more inflammation and oxidative damage, increased leukocyte infiltration, increased arthritis in mice	[45, 46]
Mouse psoriatic skin inflammation	Nrf2 deficiency	increased inflammation and psoriasis	[49]
Imiquimod-induced psoriasis mouse model	Nrf2 siRNA	reduced inflammation and psoriasis	[50]
PBMC from MS patients	Nrf2 activator	inhibited IL-17a production	[35]
Mouse EAE	Nrf2 activator	reduced EAE	[40–44]
LPS-induced arthritis in mouse	Nrf2 activator	attenuated osteoclasts	[47]
Imiquimod-induced psoriasis mouse model	Nrf2 activator	alleviated psoriasis severity	[51]

Table 1.
Summary of Nrf2 study in autoimmune disease model.

have shown that the mice exhibit increased inflammatory responses to imiquimod-induced psoriasis, as evidenced by increases in ear thickness and elevated mRNA expression of key pro-inflammatory cytokines, such as IL-6, TNFα, IL-17a, and IL-23a, in the ear tissue [49]. Moreover, topical application of Nrf2 siRNA to the ear skin resulted in reduced inflammation in mice with imiquimod-induced psoriasis [50]. DMF has been observed to enhance epidermal barrier function and alleviate the severity of psoriatic lesions [49]. Additionally, treatment with sulforaphane in mice has demonstrated efficacy in ameliorating psoriatic symptoms and protecting against cutaneous inflammation in an Nrf2-dependent manner [51]. The effect of Nrf2 modulation on autoimmune disease is summarized in **Table 1**. These findings suggest that targeting Nrf2 may serve as a potential therapeutic approach for managing psoriasis.

5. Conclusion

Overall, balancing cellular redox levels for proper metabolic processes is critical for managing Th17-dependent autoimmune diseases. Targeting Nrf2-mediated

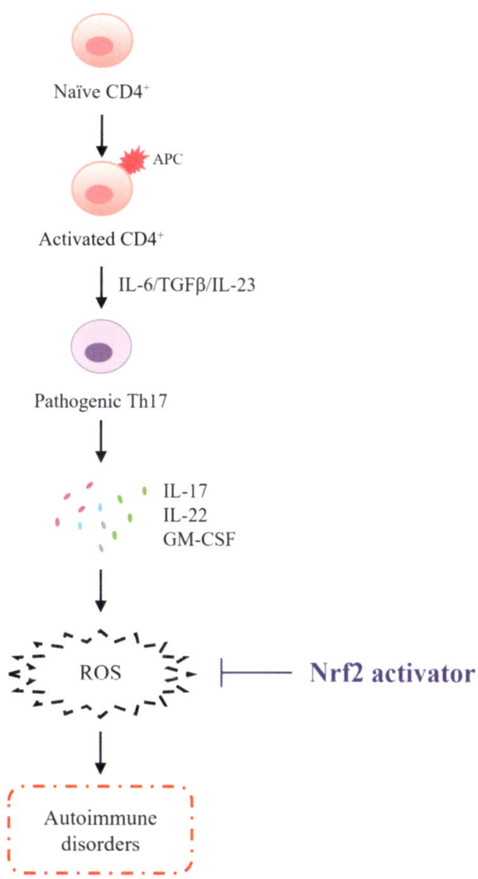

Figure 1.
Regulation of Th17 and autoimmune disease by Nrf2. Th17 cells are generated from naïve $CD4^+$ T lymphocytes upon exposure to a pathogenic environment (e.g., IL-6, TGFβ and IL-23). The differentiated Th17 cells secrete pathogenic cytokines including IL-17, IL-22, and GM-CSF, thereby facilitating the development of autoimmune disease through elevated ROS production. Activation of Nrf2 inhibits ROS and protects against autoimmune disease development.

signaling pathways offers a promising avenue for precision medicine and improving outcomes in such conditions. By modulating Nrf2 activity, it becomes possible to regulate Th17 cell differentiation and function, thereby mitigating autoimmune pathology (**Figure 1**). This approach holds considerable potential for developing novel therapeutic interventions that can effectively manage Th17-mediated autoimmune diseases.

Abbreviations

ARE	antioxidant response elements
DMF	dimethyl fumarate
EAE	experimental autoimmune encephalomyelitis
ETC	electron transport chain
GLUT	glucose transporter
HIF1α	hypoxia-inducible factor 1α
IL-17	interleukin-17
MS	multiple sclerosis
Nrf2	nuclear factor erythroid 2-related factor 2
OXPHOs	oxidative phosphorylation
PDHK1	pyruvate dehydrogenase kinase 1
PKM2	pyruvate kinase isoform 2
RA	rheumatoid arthritis
RORγt	RAR-related orphan receptor gamma
ROS	reactive oxygen species
TCA	tricarboxylic acid cycle
CDDO-TFEA	trifluoroethyl amide derivate of 2-cyano-3,12-dioxooleana-1,9(11)-dien-28-oic acid

Author details

Hongmin Wu* and Xiancai Zhong
Department of Immunology and Theranostics, Arthur Riggs Diabetes and Metabolism Research Institute, Beckman Research Institute of the City of Hope, Duarte, CA, USA

*Address all correspondence to: howu@coh.org

IntechOpen

© 2024 The Author(s). Licensee IntechOpen. This chapter is distributed under the terms of the Creative Commons Attribution License (http://creativecommons.org/licenses/by/3.0), which permits unrestricted use, distribution, and reproduction in any medium, provided the original work is properly cited.

References

[1] Lee Y, Awasthi A, Yosef N, Quintana FJ, Xiao S, Peters A, et al. Induction and molecular signature of pathogenic TH17 cells. Nature Immunology. 2012;**13**:991-999. DOI: 10.1038/ni.2416

[2] Codarri L, Gyülvészi G, Tosevski V, Hesske L, Fontana A, Magnenat L, et al. RORgammat drives production of the cytokine GM-CSF in helper T cells, which is essential for the effector phase of autoimmune neuroinflammation. Nature Immunology. 2011;**12**:560-567. DOI: 10.1038/ni.2027

[3] Sun L, Fu J, Zhou Y. Metabolism controls the balance of Th17/T-regulatory cells. Frontiers in Immunology. 2017;**8**:1632. DOI: 10.3389/fimmu.2017.01632

[4] Papadopoulou G, Xanthou G. Metabolic rewiring: A new master of Th17 cell plasticity and heterogeneity. The FEBS Journal. 2022;**289**:2448-2466. DOI: 10.1111/febs.15853

[5] Imboden JB, Stobo JD. Transmembrane signalling by the T cell antigen receptor. Perturbation of the T3-antigen receptor complex generates inositol phosphates and releases calcium ions from intracellular stores. The Journal of Experimental Medicine. 1985;**161**:446-456. DOI: 10.1084/jem.161.3.446

[6] Murphy AN, Fiskum G, Beal MF. Mitochondria in neurodegeneration: Bioenergetic function in cell life and death. Journal of Cerebral Blood Flow and Metabolism. 1999;**19**:231-245. DOI: 10.1097/00004647-199903000-00001

[7] Yarosz EL, Chang CH. The role of reactive oxygen species in regulating T cell-mediated immunity and disease. Immune Network. 2018;**18**:e14. DOI: 10.4110/in.2018.18.e14

[8] Jaramillo MC, Zhang DD. The emerging role of the Nrf2-Keap1 signaling pathway in cancer. Genes & Development. 2013;**27**:2179-2191. DOI: 10.1101/gad.225680.113

[9] Pant T, Uche N, Juric M, Zielonka J, Bai X. Regulation of immunomodulatory networks by Nrf2-activation in immune cells: Redox control and therapeutic potential in inflammatory diseases. Redox Biology. 2024;**70**:103077. DOI: 10.1016/j.redox.2024.103077

[10] Wu WJ, Jia WW, Liu XH, Pan LL, Zhang QY, Yang D, et al. S-propargyl-cysteine attenuates inflammatory response in rheumatoid arthritis by modulating the Nrf 2-ARE signaling pathway. Redox Biology. 2016;**10**:157-167. DOI: 10.1016/j.redox.2016.08.011

[11] Omenetti S, Bussi C, Metidji A, Iseppon A, Lee S, Tolaini M, et al. The intestine harbors functionally distinct homeostatic tissue-resident and inflammatory Th17 cells. Immunity. 2019;**51**:77-89.e6. DOI: 10.1016/j.immuni.2019.05.004

[12] Zhang D, Jin W, Wu R, Li J, Park SA, Tu E, et al. High glucose intake exacerbates autoimmunity through reactive-oxygen-species-mediated TGF-β cytokine activation. Immunity. 2019;**51**:671-681. DOI: 10.1016/j.immuni.2019.08.001

[13] Dang EV, Barbi J, Yang HY, Jinasena D, Yu H, Zheng Y, et al. Control of T(H)17/T(reg) balance by hypoxia-inducible factor 1. Cell. 2011;**146**:772-784. DOI: 10.1016/j.cell.2011.07.033

[14] Michalek RD, Gerriets VA, Jacobs SR, Macintyre AN, MacIver NJ, Mason EF, et al. Cutting edge: Distinct glycolytic and lipid oxidative metabolic programs are essential for effector and regulatory CD4+ T cell subsets. Journal of Immunology. 2011;**186**:3299-3303. DOI: 10.4049/jimmunol.1003613

[15] Hochrein SM, Wu H, Eckstein M, Arrigoni L, Herman JS, Schumacher F, et al. The glucose transporter GLUT3 controls T helper 17 cell responses through glycolytic-epigenetic reprogramming. Cell Metabolism. 2022;**34**:516-532. DOI: 10.1016/j.cmet.2022.02.015

[16] Chen C, Zhang W, Zhou T, Liu Q, Han C, Huang Z, et al. Vitamin B5 rewires Th17 cell metabolism via impeding PKM2 nuclear translocation. Cell Reports. 2022;**41**:111741. DOI: 10.1016/j.celrep.2022.111741

[17] Gerriets VA, Kishton RJ, Nichols AG, Macintyre AN, Inoue M, Ilkayeva O, et al. Metabolic programming and PDHK1 control CD4+ T cell subsets and inflammation. Journal of Clinical Investigation. 2015;**125**:194-207. DOI: 10.1172/JCI76012

[18] Nissler K, Petermann H, Wenz I, Brox D. Fructose 2,6-bisphosphate metabolism in Ehrlich ascites tumour cells. Journal of Cancer Research and Clinical Oncology. 1995;**121**:739-745. DOI: 10.1007/BF01213320

[19] Kosaisawe N, Sparta B, Pargett M, Teragawa CK, Albeck JG. Transient phases of OXPHOS inhibitor resistance reveal underlying metabolic heterogeneity in single cells. Cell Metabolism. 2021;**33**:649-665.e8. DOI: 10.1016/j.cmet.2021.01.014

[20] Shin B, Benavides G, Geng J, Koralov S, Hu H, Darley-Usmar V, et al. Mitochondrial oxidative phosphorylation regulates the fate decision between pathogenic Th17 and regulatory T cells. Cell Reports. 2020;**30**:1898-1909.e4. DOI: 10.1016/j.celrep.2020.01.022

[21] Franchi L, Monteleone I, Hao L, Spahr M, Zhao W, Liu X, et al. Inhibiting oxidative phosphorylation *in vivo* restrains Th17 effector responses and ameliorates murine colitis. Journal of Immunology. 2017;**198**:2735-2746. DOI: 10.4049/jimmunol.1600810

[22] Bailis W, Shyer J, Zhao J, Canaveras J, Al Khazal F, Qu R, et al. Distinct modes of mitochondrial metabolism uncouple T cell differentiation and function. Nature. 2019;**571**:403-407. DOI: 10.1038/s41586-019-1311-3

[23] Fuseini H, Cephus J, Wu P, Davis J, Contreras D, Gandhi V, et al. Erα signaling increased IL-17A production in Th17 cells by upregulating IL-23R expression, mitochondrial respiration, and proliferation. Frontiers in Immunology. 2019;**10**:2740. DOI: 10.3389/fimmu.2019.02740

[24] Zhou J, Deo BK, Hosoya K, Terasaki T, Obrosova IG, Brosius FC 3rd, et al. Increased JNK phosphorylation and oxidative stress in response to increased glucose flux through increased GLUT1 expression in rat retinal endothelial cells. Investigative Ophthalmology & Visual Science. 2005;**46**:3403-3410. DOI: 10.1167/iovs.04-1064

[25] Kussmaul L, Hirst J. The mechanism of superoxide production by NADH: Ubiquinone oxidoreductase (complex I) from bovine heart mitochondria. Proceedings of the National Academy of Sciences of the United States of America. 2006;**103**:7607-7612. DOI: 10.1073/pnas.0510977103

[26] Kamiński MM, Sauer SW, Kamiński M, Opp S, Ruppert T,

Grigaravičius P, et al. T cell activation is driven by an ADP-dependent glucokinase linking enhanced glycolysis with mitochondrial reactive oxygen species generation. Cell Reports. 2012;**2**:1300-1315. DOI: 10.1016/j.celrep.2012.10.009

[27] Nishikawa T, Edelstein D, Du XL, Yamagishi S, Matsumura T, Kaneda Y, et al. Normalizing mitochondrial superoxide production blocks three pathways of hyperglycaemic damage. Nature. 2000;**404**:787-790. DOI: 10.1038/35008121

[28] Yu T, Jhun BS, Yoon Y. High-glucose stimulation increases reactive oxygen species production through the calcium and mitogen-activated protein kinase-mediated activation of mitochondrial fission. Antioxidants & Redox Signaling. 2011;**14**:425-437. DOI: 10.1089/ars.2010.3284

[29] Martínez-Reyes I, Diebold LP, Kong H, Schieber M, Huang H, Hensley CT, et al. TCA cycle and mitochondrial membrane potential are necessary for diverse biological functions. Molecular Cell. 2016;**61**:199-209. DOI: 10.1016/j.molcel.2015.12.002

[30] Kim JW, Tchernyshyov I, Semenza GL, Dang CV. HIF-1-mediated expression of pyruvate dehydrogenase kinase: A metabolic switch required for cellular adaptation to hypoxia. Cell Metabolism. 2006;**3**:177-185. DOI: 10.1016/j.cmet.2006.02.002

[31] Tse H, Thayer T, Steele C, Cuda C, Morel L, Piganelli J, et al. NADPH oxidase deficiency regulates Th lineage commitment and modulates autoimmunity. Journal of Immunology. 2010;**185**:5247-5258. DOI: 10.4049/jimmunol.1001472

[32] Zhi L, Ustyugova IV, Chen X, Zhang Q, Wu MX. Enhanced Th17 differentiation and aggravated arthritis in IEX-1-deficient mice by mitochondrial reactive oxygen species-mediated signaling. Journal of Immunology. 2012;**189**:1639-1647. DOI: 10.4049/jimmunol.1200528

[33] Liu X, Hu Z, Zhou H. N-acetylcysteine improves inflammatory response in COPD patients by regulating Th17/Treg balance through hypoxia inducible factor-1α pathway. BioMed Research International. 2021;**2021**:6372128. DOI: 10.1155/2021/6372128

[34] Zhao M, Chen H, Ding Q, Xu X, Yu B, Huang Z. Nuclear factor erythroid 2-related factor 2 deficiency exacerbates lupus nephritis in B6/lpr mice by regulating Th17 cell function. Scientific Reports. 2016;**6**:38619. DOI: 10.1038/srep38619

[35] Lin X, Tawch S, Wong HT, Roy S, Gaudino S, Castillo P, et al. Nrf2 through aryl hydrocarbon receptor regulates IL-22 response in CD4(+) T cells. Journal of Immunology. 2021;**206**:1540-1548. DOI: 10.4049/jimmunol.1900656

[36] Noel S, Martina MN, Bandapalle S, Racusen LC, Potteti HR, Hamad AR, et al. T lymphocyte-specific activation of Nrf2 protects from AKI. Journal of the American Society of Nephrology. 2015;**26**:2989-3000. DOI: 10.1681/ASN.2014100978

[37] Dhib-Jalbut S. Pathogenesis of myelin/oligodendrocyte damage in multiple sclerosis. Neurology. 2007;**68**:S13-S21. DOI: 10.1212/01.wnl.0000275228.13012.7b

[38] Hayashi M, Takai J, Yu L, Motohashi H, Moriguchi T, Yamamoto M. Whole-body *in vivo* monitoring of inflammatory diseases exploiting human interleukin 6-luciferase transgenic

mice. Molecular and Cellular Biology. 2015;**35**:3590-3601. DOI: 10.1128/MCB.00506-15

[39] Johnson DA, Amirahmadi S, Ward C, Fabry Z, Johnson JA. The absence of the pro-antioxidant transcription factor Nrf2 exacerbates experimental autoimmune encephalomyelitis. Toxicological Sciences. 2010;**114**:237-246. DOI: 10.1093/toxsci/kfp274

[40] Linker RA, Lee DH, Ryan S, Van Dam AM, Conrad R, Bista P, et al. Fumaric acid esters exert neuroprotective effects in neuroinflammation via activation of the Nrf2 antioxidant pathway. Brain. 2011;**134**:678-692. DOI: 10.1093/brain/awq386

[41] Hammer A, Waschbisch A, Knippertz I, Zinser E, Berg J, Jörg S, et al. Role of nuclear factor (erythroid-derived 2)-like 2 signaling for effects of fumaric acid esters on dendritic cells. Frontiers in Immunology. 2017;**8**:1922. DOI: 10.3389/fimmu.2017.01922

[42] Pareek TK, Belkadi A, Kesavapany S, Zaremba A, Loh SL, Bai L, et al. Triterpenoid modulation of IL-17 and Nrf-2 expression ameliorates neuroinflammation and promotes remyelination in autoimmune encephalomyelitis. Scientific Reports. 2011;**1**:201. DOI: 10.1038/srep00201

[43] Kuo PC, Brown DA, Scofield BA, Yu IC, Chang FL, Wang PY, et al. 3H-1,2-dithiole-3-thione as a novel therapeutic agent for the treatment of experimental autoimmune encephalomyelitis. Brain, Behavior, and Immunity. 2016;**57**:173-186. DOI: 10.1016/j.bbi.2016.03.015

[44] Higashi C, Kawaji A, Tsuda N, Hayashi M, Saito R, Yagishita Y, et al. The novel Nrf 2 inducer TFM-735 ameliorates experimental autoimmune encephalomyelitis in mice. European Journal of Pharmacology. 2017;**802**:76-84. DOI: 10.1016/j.ejphar.2017.02.044

[45] Maicas N, Ferrándiz ML, Brines R, Ibáñez L, Cuadrado A, Koenders MI, et al. Deficiency of Nrf2 accelerates the effector phase of arthritis and aggravates joint disease. Antioxidants & Redox Signaling. 2011;**15**:889-901. DOI: 10.1089/ars.2010.3835

[46] Wruck CJ, Fragoulis A, Gurzynski A, Brandenburg LO, Kan YW, Chan K, et al. Role of oxidative stress in rheumatoid arthritis: Insights from the Nrf 2-knockout mice. Annals of the Rheumatic Diseases. 2011;**70**:844-850. DOI: 10.1136/ard.2010.132720

[47] Yamaguchi Y, Kanzaki H, Katsumata Y, Itohiya K, Fukaya S, Miyamoto Y, et al. Dimethyl fumarate inhibits osteoclasts via attenuation of reactive oxygen species signalling by augmented antioxidation. Journal of Cellular and Molecular Medicine. 2018;**22**:1138-1147. DOI: 10.1111/jcmm.13367

[48] Wu X, Tian J, Wang S. Insight into non-pathogenic Th17 cells in autoimmune diseases. Frontiers in Immunology. 2018;**9**:1112. DOI: 10.3389/fimmu.2018.01112

[49] Ogawa T, Ishitsuka Y, Inoue S, Nakamura Y, Saito A, Okiyama N, et al. Nuclear factor erythroid 2-related factor 2 (Nrf 2) regulates epidermal keratinization under psoriatic skin inflammation. The American Journal of Pathology. 2020;**190**:577-585. DOI: 10.1016/j.ajpath.2019.10.022

[50] Yang L, Fan X, Cui T, Dang E, Wang G. Nrf2 promotes keratinocyte proliferation in psoriasis through up-regulation of keratin 6, keratin 16, and keratin 17. Journal of Investigative

Dermatology. 2017;**137**:2168-2176. DOI: 10.1016/j.jid.2017.05.015

[51] Ma C, Gu C, Lian P, Wazir J, Lu R, Ruan B, et al. Sulforaphane alleviates psoriasis by enhancing antioxidant defense through KEAP1-NRF2 pathway activation and attenuating inflammatory signaling. Cell Death & Disease. 2023;**14**:768. DOI: 10.1038/s41419-023-06234-9

Chapter 6

NRF2 in Neurological Disorders: A Molecular Beacon for Therapeutics

Neeraj Verma

Abstract

Neurological disorders represent a complex spectrum of conditions, ranging from neurodegenerative diseases to acute injuries, each posing unique challenges to treatment. This chapter delves into the emerging role of the NRF2 transcription factor as a molecular beacon with therapeutic implications in the realm of neurological disorders. NRF2, a master regulator of cellular defense mechanisms, orchestrates antioxidant responses and mitigates oxidative stress—processes intricately linked to neuroprotection. The chapter explores the intricate interplay between NRF2 and neurological pathologies, emphasizing its influence on the progression of conditions such as Alzheimer's, Parkinson's, and ischemic stroke. By dissecting the molecular pathways through which NRF2 modulates inflammation, oxidative damage, and apoptosis in the nervous system, we gain insights into potential therapeutic strategies. Exciting research on NRF2 activators, both natural and synthetic, offers promising avenues for drug development. Furthermore, the chapter navigates through preclinical and clinical studies, highlighting the potential of NRF2-targeted interventions in preserving neuronal function and promoting recovery. As we unravel the molecular intricacies of NRF2 in neurological contexts, this chapter serves as a guide to understanding its therapeutic potential, paving the way for innovative strategies aimed at mitigating the burden of neurological disorders.

Keywords: NRF2 transcription factor, neurological disorders, oxidative stress, therapeutic implications, molecular pathways

1. Introduction

The Nuclear factor erythroid 2-related factor 2 (NRF2) is a known transcription factor encoded in humans by the NFE2L2 gene that plays an important role in cellular protection against oxidative stress (OS) by activating genes that protect cells. Oxidative stress (OS) is a condition of an imbalance of production of reactive oxygen species (ROS) and nitrogen species (RNS) ratio and the ability of cells to protect them with antioxidants. Excessive production of reactive oxygen species (ROS) leads to changes in important cellular macromolecules, including proteins, DNA, and lipids. Many studies have demonstrated an association between oxidative stress (OS) and various neurodegenerative diseases [1, 2], including Alzheimer's disease (AD) [3], Huntington's disease (HD) [4], Parkinson's disease (PD) [5], multiple sclerosis (MS) [6], and amyotrophic lateral sclerosis (ALS) [7]. ROS/RNS are generated from different sources within cellular compartments, either from normal physiological processes or because of exposure to damage or disease in the environment [8]. Superoxide (O_2^-) is the most severe form of reactive oxygen species (ROS) and is produced by the single electron reduction of O_2 in mitochondria. Superoxide can also be produced by a group of NADPH oxidases (NOX) using oxygen and NADPH as substrates, where superoxide is rapidly removed [9]. Hydroxyl radicals (•OH) are important by-products of mitochondrial oxidative phosphorylation. These chemicals exhibit instability, high reactivity, and produce more reactive aldehydes through the process of membrane lipid peroxidation (LP), ultimately leading to cell death. Additionally, superoxide dismutase 1 (SOD1) rapidly produces reactive oxygen species (ROS) hydrogen peroxide (H_2O_2) in the cytoplasm, while extracellular H_2O_2 is produced by extracellular superoxide dismutase 3 (SOD3). Consequently, H_2O_2 may be a product of beta-oxidation of fatty acids by cytochrome P450 enzymes. When H_2O_2 is exposed to Fe^{2+} or Cu^+ (Fenton reaction), it can transform into hydroxyl radicals, although they are stable and less active [10]. The main reactive nitrogen species (RNS) is $ONOO^-$, which rapidly decomposes into HO•, nitrogen dioxide radicals (NO_2•), and nitro cations (NO_2^+) [11]. This drug is neurotoxic. Targeted therapy for neurodegenerative diseases involves addressing the various mechanisms involved in the formation of the operating system (OS) and imbalances in the endogenous cellular protection of neuronal cells [12, 13].

NRF2 plays an important role in the development and activation of blood cells, a group of enzymes responsible for metabolizing drugs [14]. Proteins whose expression increases as a result of NRF2 signaling include heme oxygenase-1 (HO-1), SOD1, catalase, and enzymes involved in glutathione (GSH) metabolism, such as glutathione S-transferase (GST), glutathione cysteine. Ligase modifier subunit and glutathione cysteine ligase catalytic subunit (GCLC) [15, 16].

The ability of NRF2 to regulate central metabolism and mitochondrial function suggests that NRF2 activation is a good and effective strategy for the treatment of neurodegenerative diseases [17]. This review explores the importance of oxidative stress in neurodegenerative diseases and the benefits of targeting the NRF2 pathway as an antioxidant and cytoprotective response. It also proposes new therapies designed to treat oxidative damage and neuroinflammation by activating the NRF2 signaling pathway.

2. Recent findings in specific disorders

2.1 Alzheimer's disease

Forgetfulness and cognitive impairment are hallmarks of Alzheimer's, the most common type of dementia. AD is an age related neurodegenerative disease. Without changes in treatment, the number of AD patients in the United States is expected to increase from currently more than 5 million to approximately 13.8 million by 2050 [18]. AD is characterized by a variety of diseases, including brain atrophy due to neuronal and synaptic loss, senile plaques composed primarily of fibrillar amyloid (Aβ) peptide, and neuronal plaques composed of hyperphosphorylated tau, a cytoskeletal protein fiber tangle (NFT). β- and γ-secretase cleave amyloid precursor protein (APP) to form Aβ. Small oligomers of self-assembled peptides are thought to contribute to brain damage in AD [19]. Aging is an important risk factor for AD, and although most cases of AD are unrelated, a small proportion have a familial basis. Early onset familial disease is associated with mutations in APP, presenilin 1 (PS1) and presenilin 2 [20].

In addition to plaques and NFTs, AD brains also feature reactive gliosis, mitochondrial dysfunction, and oxidative damage to proteins and lipids [21, 36].

2.1.1 NRF2-related genes in AD cells

Early studies have shown that the NAD(P)H:quinone oxidoreductase 1 (NQO1) content in AD tissues is higher than in control tissue [22–24] A. The same is true for both was also observed for oxygenase 1 (HO-1), glutathione reductase, and glutathione peroxidase [25–27]. Additionally, AD brains have higher levels of p62, HO-1, and glutamate cysteine ligase modifier subunit (GCLM). In addition, Tanji and colleagues [28] found co-localization of Keap1 with NFTs, evidence of interaction between them, and the presence of both p62 and Keap1 in both the soluble and insoluble protein fractions in AD cells. According to a study examining NRF2 by immunohistochemistry, NRF2 was detected in the cytoplasm and nucleus of healthy hippocampal neurons, but only in the cytoplasm in AD [29]. This suggests that NRF2 activation is lower in AD cells than in previous studies. In fact, studies have shown that some of the same NRF2-dependent genes described previously are downregulated or unchanged in AD [30–32]. These conflicting results are not surprising and may have much to do with the stage of disease at the time the tissues were collected. Additionally, according to our unpublished findings, most NRF2-dependent genes and proteins and glutathione (GSH) levels in the brain and spinal cord are NRF2-dependent. It is not reduced in −/− mice (**Figure 1**) [33, 34].

2.2 Parkinson's disease

Parkinson's disease (PD) is the most common neurodegenerative disease. It is characterized by a variety of non-motor neuropsychiatric problems including motor symptoms such as tremor, bradykinesia, rigidity, and postural instability [35]. Loss of dopaminergic neurons in the substantia nigra pars compacta (SN), which reduces the number of dopamine nerve terminals projecting to the striatum, is one of the symptoms of Parkinson's disease (PD). Another notable feature is the formation of intracellular inclusions or Lewy bodies (LB) [36–38] composed of ubiquitin-linked α-synuclein (SYN). Similar to AD, PD is generally rare, with familial onset accounting for approximately 15% of cases [39]. α-SYN, parkin, leucine-rich repeat kinase 2 (LRRK2), putative PTEN-induced kinase 1 (PINK1), and DJ1 are among many genes associated with familial Parkinson's disease (PD) [40]. Just as Alzheimer's has both sporadic and familial features, Parkinson's disease tends to run in families. Therefore, the majority of familial PD and AD are early-onset, progressive diseases.

Studies have shown that NQO1 expression is increased in astrocytes, endothelial cells, and dopaminergic neurons in people with Parkinson's disease [41]. HO-1 and peroxidase in the substantia nigra of Parkinson's disease patients show a similar pattern [42]. Immunostaining demonstrated the presence of NRF2.

2.2.1 NRF2-related genes in PD cells

Nuclear localization of NRF2 was observed in dopaminergic neurons in the Parkinson's disease brain [29]. Oxidative stress, mitochondrial dysfunction, protein carbonylation, and oxidative damage to substantia nigra DNA have been implicated in Parkinson's disease [43, 44]. Glutathione concentrations are reduced in the brain of patients with Parkinson's disease [45, 46]. In addition, Keap1 colocalizes with

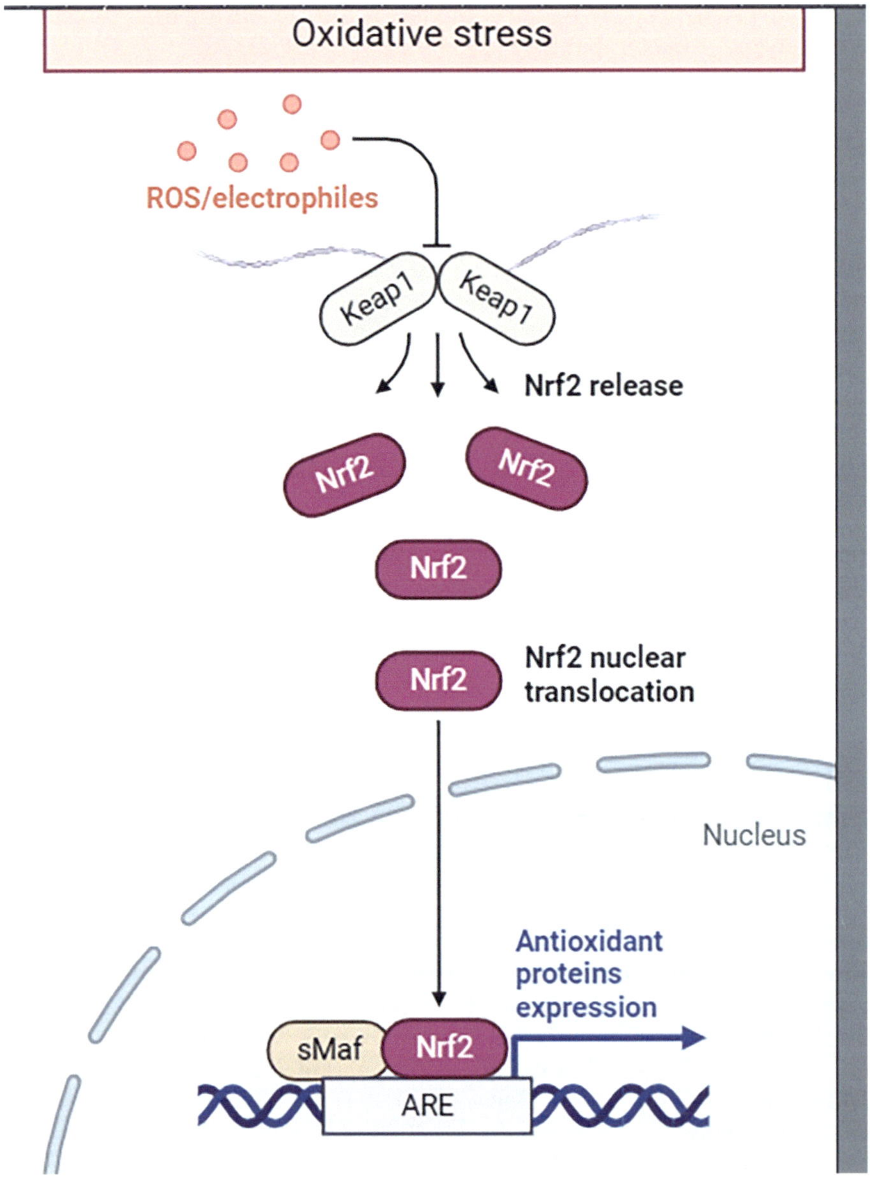

Figure 1.
The Keap1-NRF2 pathway is a signaling pathway involved in cellular responses to oxidative stress. The regulatory protein Keap1 regulates the activity of the transcription factor NRF2. Under conditions of oxidative stress, NRF2 is released from Keap1 and translocates to the nucleus, where it activates the expression of genes encoding antioxidant enzymes. This pathway plays an important role in cell defense against oxidative damage and redox maintenance of homeostasis.

SYN and p62 in Lewy bodies in the Parkinson's disease brain [28]. New research on induced pluripotent stem cells (iPSCs) from patients with Parkinson's disease (PD) and control subjects provides further evidence of the role of NRF2 in Parkinson's disease. Neurons derived from Parkinson's disease-induced pluripotent stem cells (PD iPSCs) show decreased glutathione (GSH), increased oxidative stress, increased NRF2 activation, and increased NQO1 levels [47].

2.3 Amyotrophic lateral sclerosis

Amyotrophic lateral sclerosis (ALS) is a serious motor disease in adults and is characterized by the deterioration of slow motor neurons [48]. Motor neurons have intracellular inclusions that are ubiquitin-positive, hyaline, and have a twisted appearance [49]. These include SOD1, which colocalizes with p62 [50]. ALS has also been shown to have TDP-43 (TAR DNA binding protein) and FUS-positive inclusions [51, 52]. Like AD and PD, most ALS patients have a variant form (sALS); Approximately 10–15% of cases are associated with a familial variant (fALS). The first genetic mutation found in familial amyotrophic lateral sclerosis (fALS) was in copper/zinc superoxide dismutase 1 (SOD1), which causes disease by increasing its activity [53]. More than 150 mutations have been identified in the SOD1 gene. Mice with an extreme hSOD1 mutant often have an ALS-like phenotype [54, 55]. The specific molecular mechanism responsible for the increase in dysfunction is currently unknown. However, toxicity to motor neurons requires expression of mutant SOD1 in both non-neuronal cells and motor neurons [56]. Over the last few years, researchers have discovered 24 additional genes associated with amyotrophic lateral sclerosis (sALS) and familial amyotrophic lateral sclerosis (fALS) [57]. One of these genes is p62, which directly affects Keap1/NRF2 and autophagy [58].

2.3.1 NRF2-related genes in ALS cells

The presence of NRF2 and Keap1 in the primary motor cortex and spinal cord of ALS patients was investigated [59]. NRF2 mRNA and protein levels were reduced in ALS patient tissues compared with control tissues, but Keap1 mRNA and protein levels were unchanged. NRF2 dependent genes were not analyzed. Furthermore, Keap1 has been shown to co-localize with intracellular inclusions in post-mortem ALS spinal motor neurons [28].

2.4 Huntington's disease

Huntington's disease (HD) is a neurological disease caused by the expansion of CAG repeats in the gene responsible for producing the Htt protein [60]. This leads to elongation of the N-terminal glutamine residue, and the intensity and duration of the disease are related to the length of the polyglutamine repeat. In addition, structural changes in mutant Htt lead to increased self-aggregation, which is associated with the appearance of intracellular inclusions [61]. Huntington's disease (HD) is defined by pathological degeneration of the neostriatal regions (caudate and putamen) and cerebral cortex. This degeneration is thought to be the main cause of mobility, cognitive impairment, and psychiatric symptoms that worsen as the disease progresses. Oxidative stress due to mitochondrial dysfunction has been shown to be responsible in human patients [62]. According to data obtained from postmortem brain examination, dysfunction of mitochondrial complex II, III, and IV was detected in the striatum of deceased Huntington's disease patients [63, 64].

2.4.1 NRF2-related genes in HD cells

Surprisingly, very little work has been done on HD tissues and NRF2. Only one study investigated the presence of NRF2 or NRF2-dependent genes in post-mortem HD cells [65]. This study demonstrated HO-1 immunohistochemical staining in HD cells.

2.5 Multiple sclerosis

Approximately 400,000 people in the United States and more than 2.3 million people worldwide are thought to be affected by multiple sclerosis (MS). Multiple sclerosis (MS) is a chronic disease characterized by inflammation of the nervous system. It begins with the activation of CD4+ T lymphocytes in the peripheral zone, which then cross the blood-brain barrier and initiate an autoimmune attack on myelin and oligodendrocytes in the central nervous system (CNS) [66]. Neurological disorders occur in a variety of conditions, including vision loss, coordination problems, fatigue, hearing loss, cognitive impairment, and bladder and bowel problems. Multiple sclerosis (MS) disease occurs as a result of the destruction of the immune system, and autoreactive CD4+ Th1 and Th17 cells (myelin sheath), which are the protective covering of the brain, emerge and myelin-producing cells (oligodendrocytes) emerge. This attack is also accomplished by infiltrating macrophages and resident microglia. As a result, nerve cells decrease and the number of stellate cells (a type of glial cell) increases in response to the damage. Activated microglia and stellate cells secrete inflammatory substances such as cytokines and chemokines, as well as reactive oxygen species and nitrogen. These drugs exacerbate oxidative stress and further exacerbate the deleterious tissue damage associated with neuroinflammation, leading to disease progression [67, 68]. Currently, the FDA has approved eight drugs to treat relapsing-remitting MS, the most common form of MS, affecting 87 percent of people with MS. The main purpose of this medication is to prevent the immune system from crossing the blood-brain barrier. However, due to the complexity and diversity of diseases, the evolution of treatment over time, and the general belief that oxidative stress plays a role in disease, further research is required to improve and/or supplement existing treatments [69].

2.5.1 NRF2-related genes in MS cells

Damage to brain tissue in multiple sclerosis causes extensive oxidative damage resulting from the production of reactive oxygen species and nitrogen species resulting from inflammation. Antioxidants and enzymes, as well as NRF2 protein levels, have been shown to serve as markers of cellular stress in active disease compared to normal white matter (NAWM) and normal tissue in the brain that are not neuronally controlled [70–72]. Therefore, initiation of NRF2-mediated antioxidant response may indicate the existence of cellular defense mechanisms that prevent damage from oxidative stress caused by long-term inflammation. Previous studies have shown that NQO1, an enzyme putatively regulated by NRF2, is increased in multiple sclerosis (MS). This stimulation was particularly evident in hypertrophic stellate cells and myelin-laden macrophages. Oligodendrocytes sometimes show NQO1 immunoreactivity, but neuronal NQO1 labeling is rare [70]. Additional studies have shown that the disease in multiple sclerosis (MS) exhibits strong resistance to markers of oxidative stress, including 4-HNE, 8-hydroxydeoxyguanosine, and nitrotyrosine. It has also been found that these active bacteria contain antioxidant enzymes such as HO-1, SOD1, SOD2, and catalase. Like NQO1 manifestations, this staining is seen in hypertrophic stellate cells and myelin-rich macrophages. Rare staining was observed in NAWM. Upregulation of NRF2 in MS lesions can be seen in the nuclei and cytoplasm of invasive macrophages and reactive stellate cells, with little expression in oligodendrocytes around active lesions [72].

3. Therapeutic scope of the activation or inhibition of the transcriptional factor NRF2 in neurological disorders

NRF2 activation mitigates multiple pathogenic processes involved in these neurodegenerative disorders through upregulation of antioxidant defenses, inhibition of inflammation, improvement of mitochondrial function, and maintenance of protein homeostasis.

3.1 Oxidative stress and Nrf2

NRF2 controls the production of many antioxidant enzymes and proteins that provide protection to cells. Studies show that enhancing NRF2-dependent antioxidant activities may enhance brain functions in illness models. The synaptic degeneration and neuronal death in Alzheimer's disease are a result of various disrupted processes such as heightened oxidative stress, persistent inflammation, impaired mitochondrial function, buildup of amyloid-β (Aβ) 1–42 peptides from the amyloid precursor protein (APP), proteasome inhibition, mutations in APP, presenilin-1, and presenilin-2 genes, and accumulation of hyper-phosphorylated tau protein [73]. Oxidative stress is partly caused by oligomeric Aβ peptides and occurs before other metabolic changes, leading to persistent inflammation due to oxidative damage.

3.2 Neuroinflammation and Nrf2

It is a feature of neurodegenerative diseases and is caused primarily by activated microglia (inhabitants of the immune system). When activated, microglia release cytotoxic reactive oxygen species and nitrogen, which can affect neurons, especially in cases of excessive microglial activation and dysregulation. Additionally, the presence of dead or injured neurons leads to activation of microglia (known as reactive microgliosis) through recognition receptors such as Toll-like receptors. This causes a continuous cycle of neuronal cell death. The density of microglia in the adult brain varies by region and can range between 0.5 and 16.6%. They are usually found in a dormant state with branching-like patterns in the gray matter. Regions with the most cells include the hippocampus, olfactory telencephalon, basal ganglia, and substantia nigra [74, 75]. Therefore, it is not surprising that certain regions of the brain can be affected by microgliosis, as seen in many neurodegenerative diseases, including Alzheimer's disease (AD), Parkinson's disease (PD), Huntington's disease (HD), and amyotrophic lateral multiple sclerosis (ALS) [76]. The main source of reactive chemicals produced by microglia is the activation of the NADPH oxidase (NOX) system, which produces more superoxide anions ($\bullet O_2^-$). The NOX enzyme complex in the cell membrane normally enhances the elimination of pathogens when triggered by different stimuli such as pathogen-associated molecular patterns (PAMPs), inflammatory cytokines, and neurotoxins. It also causes changes in cell shape and movement and regulates gene activity. Sustained activation of NOX leads to oxidation of lipids and other macromolecules and causes apoptosis, ultimately leading to neuronal cell death and degeneration. Similarly, enzyme-inducible nitric oxide synthase (iNOS) is stimulated, and its activation produces nitric oxide (NO•), similar to $\bullet O_2^-$ which can produce neurotoxic effects when consistently present at high concentrations. An increase in the level of NO• can disrupt the balance of redox homeostasis, and NO• can also react with $\bullet O_2^-$, leading to the formation of more harmful compounds called peroxynitrite ($ONNO^-$) [50]. At rest, microglia constantly monitor its environment for bacteria, abnormal cells, or cellular debris. They also help reorganize and maintain synapses. When activated by risk, microglia become activated and subsequently release anti-inflammatory substances such as TNF, interleukin (IL)-1, IL-6, IL-12, and IL-23. This phase of microglial activation is often referred to as the classical proinflammatory phase and is referred to as M1. M2 is the stage after the M1 stage and is characterized by a long period and immunity or change. During this time, microglia are activated in response to inflammatory cytokines, including IL-4, IL-10, and IL-13. The body then produces the same cytokines while stimulating genes involved in wound healing, including arginase 1 (ARG1), mannose receptor (MMC), and chitinase-like 3 in domain 1 (FIZZ1)-1 (Ym1 in mice). In fact, the antioxidant properties of microglia maintain cellular homeostasis and keep microglia constantly alert. NF-kappaB (NF-kß) is a master regulator of the brain's response to infection, the environment, and cellular damage. The redox state of microglia plays an important role in regulating the nuclear level of NF-kß as well as the flux of NF-kß kinases [74]. Microglia function well when the body's antioxidant defense system maintains oxidative stress (OS) at normal levels. However, when OS or neuroinflammation (NS) persists, the NF-kß signaling pathway remains active. This causes more inflammation in microglia, making them more active and affecting the nervous system. Transcription of NRF2 is responsible for the activation of antioxidant genes and maintenance of redox homeostasis. It is a master regulator of microglia fate, protecting microglia from oxidative stress and preventing hyperactivation of the M1 phenotype. Activation of NRF2 specifically leads to increased expression of genes involved in various processes, including elimination of oxidative stress (SOD3, GPx, and Prx), production of reducing agents (GCLM/C or NADPH), and regeneration of cofactors. Proteins (GR, Trx-TrxR, Prx/thioredoxin), transport of redox molecules (xCT), metal chelation (MT1/2 or ferritin), and antioxidant production enzymes (HO-1, NQO1 and Trx). In addition, NRF2 activation also inhibits the expression of the Trx inhibitor TXNIP. Many studies have identified NRF2 as a useful target for the treatment of neurological diseases associated with neuronal protection secondary to neurodegenerative pathologies. In one study, mice lacking the NRF2 gene (NRF2–/–) were injected with lipopolysaccharide (LPS), a substance known to cause inflammation. These animals show sensitivity to lipopolysaccharide-induced neuroinflammation, especially in the hippocampal region. NRF2–/– mice have protein and mRNA levels of the microglial marker F4/80 in the hippocampal gyrus compared to WT animals. In addition, the inflammatory markers iNOS, IL-6, and TNF-α were also improved. Additionally, SFN-treated mice showed a two- to threefold increase in HO-1 levels, decreased microglia numbers, and decreased production of specific markers in response to LPS [77]. Mice treated with SFN after spinal cord injury showed an increase in NRF2 and GCL levels as well as a decrease in the cytokines IL-1ß and TNF-α. As a result, these mice had decreased volume and improved coordination [78]. In a mouse model of Parkinson's disease (PD), MPTP injection into animals lacking the NRF2 gene (NRF2–/–) showed strong oxidative stress (OS). These mice had significantly more impaired dopaminergic activity in the basal ganglia than control mice with normal NRF2 genes (WT littermate controls). Although there was no significant difference in the number of Cd11b-positive/CD45 advanced cells between WT and NRF2–/– mice (indicating peripheral macrophage infiltration), NRF2–/– mice showed more astrocyte hyperplasia and microgliosis. This is determined by higher expression and protein expression of GFAP and F4/80 [79]. This study confirmed previous studies showing that NRF2–/– mice exposed to MPTP showed increased neuroinflammation caused by activation of astrocytes (GFAP) and microglia (Iba-1) [80]. In addition, levels of inflammatory markers associated with microglial activation (such as COX2, iNOS, IL-6, and TNF-α) were also increased in the M1 state. In contrast, markers associated with M2 status microglia, such as FIZZ-1, YM-1, arginase-1, and IL-4, were found to be decreased. Findings indicate that NRF2 plays an important role in regulating neuroinflammation after neurotoxic injury [78]. A mouse model of tauopathy was induced in the mouse hippocampus by stereotaxic injection of an adeno-disrupting vector expressing TAUP301L. In this model, NRF2−/− animals showed high levels of microgliosis and astrocytosis. Validation included measuring TNF and IL-6 mRNA levels in the hippocampus and immunohistochemical staining for Iba-1 and GFAP. In contrast to WT mice, NRF2−/− animals did not show an increase in hippocampal HO-1 and GCLC mRNA levels and did not show cellular staining for HO-1. These findings suggest that TAUP301L-induced disease activates the NRF2 signaling pathway as a regulatory response. In line with these observations and previous studies showing that neuronal production of the chemokine fractalkine (CX3CL1) affects TAU-mediated neuroinflammation, the authors of this study showed that CX3CL1 secreted by injured neurons was exclusively associated with an increase in microglia. Receptors are found in cells. This interaction activates the NRF2-ARE signaling pathway and reduces the incidence of TAUP301L-induced microglial proliferation. Subsequent studies were supported by the finding that mice lacking CX3CR1 had increased hippocampal microgliosis in response to TAUP301L compared with normal controls. Additionally, NRF2-dependent HO-1 immunohistochemical staining was absent in these animals. So, when neurons are injured, they produce a substance called CX3CL1, which then interacts with microglia through the CX3CR1 receptor. This interaction activates the NRF2-ARE pathway, which reduces the neuroinflammatory response to injury [81].

In a study with co-culture experiments using microglial conditioned media and mouse primary neurons, degenerating neurons exposed to neurotoxic oligomeric amyloid ß (oAß) were found to induce microglia to release milk fat globule-EGF factor 8. It was found to emit signal (MFG-E8). Release of MFG-E8 increases the phagocytic activity of microglia. Additionally, microglia treated with MFG-E8 was found to activate the NRF2 signaling pathway and HO-1. However, when examining neuronal-microglia co-cultures, retention of MFG-E8 was counteracted by application of the HO-1 inhibitor tin (IV)-mesoporphyrin IX dichloride (SnMP). This suggests that increased HO-1 in microglia is important for protecting neurons from toxicity [82]. These studies suggest that message exchange between neurons and microglia and stimulation of the NRF2 signaling pathway may be a new and effective way to protect neurons, especially in the context of neurodegenerative diseases.

3.3 Mitochondrial function and Nrf2

NRF2 controls cellular defense by adjusting mitochondrial function. NRF2 activation offsets the generation of reactive oxygen species by mitochondria and protects against toxins generated by mitochondria. NRF2 function is reduced in mitochondrial illnesses such as Alzheimer's disease (AD), Parkinson's disease (PD), and Friedreich's ataxia (FRDA). Research conducted with isolated mitochondria and cultured cells has shown that a lack of NRF2 results in reduced mitochondrial fatty acid oxidation, respiration, and ATP generation. Chemical substances that activate NRF2 help maintain mitochondrial health by either stimulating mitophagy or providing protection against oxidative stress-induced damage to the mitochondrial permeability transition pore [83].

4. Association between NFE2L2 and KEAP1 haplotypes and neurodegenerative disorders

Genomic studies were performed to determine whether there were genetic differences in the genes encoding NRF2 and its repressor protein Keap1 (called NFE2L2 and KEAP1, respectively). As discussed in this review, this is based on significant evidence that the neurodegenerative effects of OS/NS in diseases such as PD, AD, and ALS can be inhibited or prevented by increasing NRF2 activity. The aim is to investigate whether genetics plays a role in the development or susceptibility to these diseases. Based on analysis of published SNPs and other genetic variants, the mutation frequency of human NFE2L2 was estimated to be 1 in every 72 base pairs. This suggests that genes may have many variations, called polymorphisms [84]. Von Otter et al. Initial data analysis in patients with Parkinson's disease (PD) identified a protective NFE2L2 haplotype, GAGCAAAA. This haplotype contains a variant in the NRF2 promoter that is active in Swedish and Polish samples and is associated with high activity [85]. In a Swedish study, the presence of GAAAA haplotype alleles was associated with a higher estimated age of onset (AAO) of Parkinson's disease (PD) by approximately 5 years per allele ($p = 0.001$). In the Polish study, individuals with the protective entire haplotype GAGCAAAA had a lower risk of developing PD; There was a difference of 0.4 and 0.2 for heterozygous and homozygous carriers, respectively ($p = 2 \times 10^{-6}$). Additionally, studies in Sweden found that GAGGG and GAAAG haplotypes were associated with an increased risk of Parkinson's disease, with odds of 2.4 and 3.7 for each allele, respectively. Importantly, the promoter haplotype AGA tended to increase the risk of PD, whereas the entire promoter haplotype GAGAAGGG showed an overall risk (difference between each haplotype allele was 2.8). However, studies have shown that the entire haplotype allele GAGCAAAAA causes a delay in the age of onset of Parkinson's disease for all haplotype alleles. Neither the Swedish nor the Polish samples found a significant effect on Parkinson's disease risk for specific SNPs alone. Genetic polymorphisms of KEAP1 were also examined. However, there was no association with age at onset (AAO) or likelihood of Parkinson's disease [85]. A recent meta-analysis investigated the genetic association of NFE2L2 variants encoding NRF2 with Parkinson's disease (PD). The review included the literature mentioned above, as well as four new independent European patient-controlled studies from Malta, Germany and Italy, and a second Swedish study. The results of this analysis showed an association between the GAGCAAAAA haplotype, including the high-activity promoter haplotype AGC, and reduced risk of PD (odds ratio (OR) of none per allele 0.8, $p = 0.012$). Additionally, this haplotype is associated with delayed onset of PD (+1.1 years per allele, $p = 0.048$) [84]. The results of this study confirm the first haplotype study linking genetic variation in the NFE2L2 gene to the development of Parkinson's disease. This latest study also reported four single nucleic acid polymorphisms (SNPs) found in haplotypes associated with age of onset (AAO) in Parkinson's disease (PD). Three of the SNPs (rs7557529 G > A, rs35652124 A > G, rs2886161 A > G) were found to accelerate the onset of PD and each allele was associated with a reduction in AAO at 1.0, 1.1, and 1.2 years, respectively. The P values of these relationships were 0.042, 0.045, and 0.021, respectively. On the other hand, the fourth SNP (rs1806649 G > A) was shown to delay the onset of PD, with each allele being associated with an increase in AAO at 1.2 years. It is worth noting that the single nucleotide polymorphism (SNP) rs35652124 is functional and is located in the promoter region of the NFE2L2 gene [86]. Unlike the Parkinson's disease (PD) studies mentioned above, analysis of the NFE2L2 promoter in a control group of Taiwanese patients did not reveal any genetic changes. Three single nucleotide polymorphisms (SNPs) in the NFE2L2 promoter were examined: rs35652124, rs6706649, and rs6721961. These SNPs can be detected alone or in combination with a genetic haplotype associated with susceptibility to Parkinson's disease (PD). However, there was no association between these SNPs or haplotype variants and susceptibility to PD. Researchers have suggested that differences between studies in Taiwan and Europe may be due to differences in ethnic and environmental variables in different regions [87]. Previous studies using AD Swedish case-control data found no evidence supporting single SNPs or haplotypes in NFE2L2 or KEAP1 as genes that increase AD risk. However, genetic analysis in the NFE2L2 gene shows a significant association between the GAAAA haplotype and the onset of Alzheimer's disease 2 years earlier than average ($p = 0.013$). This means that some common mutations in the NFE2L2 gene may affect Alzheimer's disease and therefore the onset of the disease. In studies, the presence of NFE2L2 or KEAP1 gene mutations was not associated with Mini-Mental State Examination (MMSE) results or total tau and amyloid ß 1–42 (Aß42) levels in cerebrospinal fluid (CSF) [88]. A Swedish case study of NFE2L2 and KEAP1 in ALS disease showed that a specific NFE2L2 haplotype (GGGAC) was associated with a reduced risk of ALS (or one allele = 0.62, $p = 0.015$). Additionally, haplotypes in KEAP1 (CGC) were found to be associated with delayed onset of ALS (+3.4 years per allele, $p = 0.015$). Furthermore, the distribution of ALS revealed the existence of a specific group characterized by the NFE2L2-related haplotype GAGCAGA. This subgroup reveals three function-promoting SNPs associated with increased NRF2 protein expression. Each additional allele of this haplotype was associated with a 4.0-year delay in the onset of ALS ($p = 0.008$). Each additional allele of this haplotype was associated with a 4.0-year delay in the onset of ALS ($p = 0.008$) [89]. In a separate case-control study in Italian ALS patients, three NFE2L2 promoter polymorphisms (i.e., SNPs −653 A/G, −651 G/A, and −617 C/A) were tested. However, there was no significant difference between ALS cases and controls. The frequency of the −653 A/G promoter polymorphism was slightly higher in patients than in controls but did not reach significance. Additionally, there was no association between the NFE2L2 promoter polymorphisms tested and the oxidative stress biomarkers tested [84]. Similar studies have not been conducted in patients with Huntington's disease (HD) or multiple sclerosis (MS).

5. Conclusion

The NRF2-ARE pathway is an important target for therapeutic intervention in neurodegenerative diseases, as demonstrated by the facts presented. Current data provide the highest level of evidence for Parkinson's disease (PD), amyotrophic lateral sclerosis (ALS), and multiple sclerosis (MS). However, future trials in Alzheimer's disease (AD) and Huntington's disease (HD) are needed to better understand the importance of NRF2 in these diseases. Cell and in silico high-throughput screens have identified new drugs that activate NRF2 [90]. Effective strategies may include the identification of non-covalently linked compounds with interactions between DLG or ETGE motifs and Keap1 [91, 92]. Now the US Food and Drug Administration (FDA) has approved Biogen Idec's DMF (BG-12) for the treatment of multiple sclerosis (MS). As described previously, DIF has been evaluated in previous HD models. However, it is important to note that in previous models of the above diseases, only CDDO drugs were analyzed. Reata Pharmaceuticals developed the first CDDO molecule, bardoxolone methyl (also known as CDDO methyl ester), and initiated a Phase III study for the treatment of kidney disease. Unfortunately, this study was discontinued in 2012 due to a higher incidence of cardiovascular disease in the bardoxolone methyl group compared to the placebo group [93]. A new CDDO derivative called RTA 408 has been developed and is currently being tested in several non-neurodegenerative diseases [94–97]. A clinical trial investigating the use of RTA 408 to treat Friedreich's ataxia, a neurodegenerative disease that causes cerebellar ataxia, has recently begun. This condition is characterized by a decrease in the protein frataxin, resulting in a defect in mitochondrial respiration (ClinicalTrials.gov identifier NCT02255435). The inclusion of these future CDDO drugs in the major neurodegenerative diseases mentioned in this review may be an important first step, supported by the findings of previous CDDO history.

A new aspect of treatment strategies for this system is that some cells are more effective in protecting neurons from degeneration. Compared to the protection obtained by overexpression of NRF2 in stellate cells in models of ALS and PD, the lack of protection in vivo by overexpression of NRF2 in neurons or muscle cells suggests that not only the pharmacological target but also the cell specificity is still affected. Targetability may be important when considering NRF2-dependent therapeutic approaches.

Author details

Neeraj Verma
Hygia College of Pharmacy, Lucknow, India

*Address all correspondence to: neerajcology@gmail.com

IntechOpen

© 2024 The Author(s). Licensee IntechOpen. This chapter is distributed under the terms of the Creative Commons Attribution License (http://creativecommons.org/licenses/by/3.0), which permits unrestricted use, distribution, and reproduction in any medium, provided the original work is properly cited.

References

[1] Reynolds A, Laurie C, Mosley RL, Gendelman HE. Oxidative stress and the pathogenesis of neurodegenerative disorders. International Review of Neurobiology. 2007;**82**:297-325. DOI: 10.1016/S0074-7742(07)82016-2

[2] Seminotti B, Grings M, Tucci P, Leipnitz G, Saso L. Nuclear factor erythroid-2-related factor 2 signaling in the neuropathophysiology of inherited metabolic disorders. Frontiers in Cellular Neuroscience. 2021;**15**:785057. DOI: 10.3389/fncel.2021.785057

[3] Tejo FV, Quintanilla RA. Contribution of the NRF2 pathway on oxidative damage and mitochondrial failure in Parkinson and Alzheimer's disease. Antioxidants. 2021;**10**:1069. DOI: 10.3390/antiox10071069

[4] Moretti D, Tambone S, Cerretani M, Fezzardi P, Missineo A, Sherman LT, et al. NRF2 activation by reversible KEAP1 binding induces the antioxidant response in primary neurons and astrocytes of a Huntington's disease mouse model. Free Radical Biology & Medicine. 2021;**162**:243-254. DOI: 10.1016/j.freeradbiomed.2020.10.022

[5] Petrillo S, Schirinzi T, Di Lazzaro G, D'Amico J, Colona VL, Bertini E, et al. Systemic activation of NRF2 pathway in Parkinson's disease. Movement Disorders. 2020;**35**:180-184. DOI: 10.1002/mds.27890

[6] Fischer MT, Sharma R, Lim JL, Haider L, Frischer JM, Drexhage J, et al. NADPH oxidase expression in active multiple sclerosis lesions in relation to oxidative tissue damage and mitochondrial injury. Brain. 2012;**135**:886-899. DOI: 10.1093/brain/aws012

[7] Xiao Y, Karam C, Yi J, Zhang L, Li X, Yoon D, et al. ROS-related mitochondrial dysfunction in skeletal muscle of an ALS mouse model during the disease progression. Pharmacological Research. 2018;**138**:25-36. DOI: 10.1016/j.phrs.2018.09.019

[8] Finkel T. Signal transduction by mitochondrial oxidants. The Journal of Biological Chemistry. 2012;**287**:4434-4440. DOI: 10.1074/jbc.R111.283280

[9] Nauseef WM. Detection of superoxide anion and hydrogen peroxide production by cellular NADPH oxidases. Biochimica et Biophysica Acta - General Subjects. 2014;**1840**:757-767. DOI: 10.1016/j.bbagen.2013.10.029

[10] Rojkind M, Domínguez-Rosales JA, Nieto N, Greenwel P. Role of hydrogen peroxide and oxidative stress in healing responses. Cellular and Molecular Life Sciences. 2002;**59**:1872-1891. DOI: 10.1007/PL00012511

[11] Niedzielska E, Smaga I, Gawlik M, Moniczewski A, Stankowicz P, Pera J, et al. Oxidative stress in neurodegenerative diseases. Molecular Neurobiology. 2016;**53**:4094-4125. DOI: 10.1007/s12035-015-9337-5

[12] Wasik U, Milkiewicz M, Kempinska-Podhorodecka A, Milkiewicz P. Protection against oxidative stress mediated by the NRF2/Keap1 axis is impaired in primary biliary cholangitis. Scientific Reports. 2017;**7**:44769. DOI: 10.1038/srep44769

[13] Telkoparan-Akillilar P, Panieri E, Cevik D, Suzen S, Saso L. Therapeutic targeting of the NRF2 signaling pathway in cancer. Molecules. 2021;**26**:1417. DOI: 10.3390/molecules26051417

[14] Ma Q. Role of NRF2 in oxidative stress and toxicity. Annual Review of Pharmacology and Toxicology. 2013;**53**:401-426. DOI: 10.1146/annurev-pharmtox-011112-140320

[15] Stachurska A, Ciesla M, Kozakowska M, Wolffram S, Boesch-Saadatmandi C, Rimbach G, et al. Cross-talk between microRNAs, nuclear factor E2-related factor 2, and heme oxygenase-1 in ochratoxin A-induced toxic effects in renal proximal tubular epithelial cells. Molecular Nutrition & Food Research. 2013;**57**:504-515. DOI: 10.1002/mnfr.201200360

[16] Panieri E, Telkoparan-Akillilar P, Suzen S, Saso L. The NRF2/KEAP1 Axis in the regulation of tumor metabolism: Mechanisms and therapeutic perspectives. Biomolecules. 2020;**10**:791. DOI: 10.3390/biom10050791

[17] Esteras N, Dinkova-Kostova AT, Abramov AY. NRF2 activation in the treatment of neurodegenerative diseases: A focus on its role in mitochondrial bioenergetics and function. Biological Chemistry. 2016;**397**:383-400. DOI: 10.1515/hsz-2015-0270

[18] Alzheimer's Association. 2014 Alzheimer's disease facts and figures. Alzheimer's & Dementia. 2014;**10**(2):e47-e92. DOI: 10.1016/j.jalz.2014.02.001

[19] Klein WL, Krafft GA, Finch CE. Targeting small Abeta oligomers: The solution to an Alzheimer's disease conundrum? Trends in Neurosciences. 2001;**24**:219-224. DOI: 10.1016/s0166-2236(00)01749-5

[20] Sandbrink R, Hartmann T, Masters CL, Beyreuther K. Genes contributing to Alzheimer's disease. Molecular Psychiatry. 1996;**1**:27-40. DOI: 10.1038/sj.mp.4000236

[21] Smith MA, Richey Harris PL, Sayre LM, Beckman JS, Perry G. Widespread peroxynitrite-mediated damage in Alzheimer's disease. The Journal of Neuroscience. 1997;**17**:2653-2657. DOI: 10.1523/JNEUROSCI.17-07-02653.1997

[22] Raina AK, Templeton DJ, Deak JC, Perry G, Smith MA. Quinone reductase (NQO1), a sensitive redox indicator, is increased in Alzheimer's disease. Redox Report. 1999;**4**:23-27. DOI: 10.1179/135100099101534890

[23] Wang Y, Santa-Cruz K, DeCarli C, Johnson JA. NAD(P)H:Quinone oxidoreductase activity is increased in hippocampal pyramidal neurons of patients with Alzheimer's disease. Neurobiology of Aging. 2000;**21**:525-531. DOI: 10.1016/s0197-4580(00)00126-0

[24] SantaCruz KS, Yazlovitskaya E, Collins J, Johnson J, DeCarli C. Regional NAD(P)H:Quinone oxidoreductase activity in Alzheimer's disease. Neurobiology of Aging. 2004;**25**:63-69. DOI: 10.1016/S0197-4580(03)00059-8

[25] Schipper HM, Cisse S, Stopa EG. Expression of heme oxygenase-1 in the senescent and Alzheimer-diseased brain. Annals of Neurology. 1995;**37**:758-768. DOI: 10.1002/ana.410370612

[26] Schipper HM, Chertkow H, Mehindate K, Frankel D, Melmed C, Bergman H. Evaluation of heme oxygenase-1 as a systemic biological marker of sporadic AD. Neurology. 2000;**54**:1297-1304. DOI: 10.1212/WNL.54.6.1297

[27] Aksenov MY, Markesbery WR. Changes in thiol content and expression of glutathione redox system genes in the hippocampus and cerebellum in Alzheimer's disease. Neuroscience

Letters. 2001;**302**:141-145. DOI: 10.1016/s0304-3940(01)01537-4

[28] Tanji K, Maruyama A, Odagiri S, Mori F, Itoh K, Kakita A, et al. Keap1 is localized in neuronal and glial cytoplasmic inclusions in various neurodegenerative diseases. Journal of Neuropathology and Experimental Neurology. 2013;**72**:18-28. DOI: 10.1097/NEN.0b013e31827b1221

[29] Ramsey CP, Glass CA, Montgomery MB, Lindl KA, Ritson GP, Chia LA, et al. Expression of NRF2 in neurodegenerative diseases. Journal of Neuropathology and Experimental Neurology. 2007;**66**:75-85. DOI: 10.1097/nen.0b013e31802f11bf

[30] Lovell MA, Xie C, Markesbery WR. Decreased glutathione transferase activity in brain and ventricular fluid in Alzheimer's disease. Neurology. 1998;**51**:1562-1566. DOI: 10.1212/WNL.51.6.1562

[31] Ansari MA, Scheff SW. Oxidative stress in the progression of Alzheimer disease in the frontal cortex. Journal of Neuropathology and Experimental Neurology. 2010;**69**:155-167. DOI: 10.1097/NEN.0b013e3181cb5af4

[32] Saharan S, Mandal PK. The emerging role of glutathione in Alzheimer's disease. Journal of Alzheimer's Disease. 2014;**40**:519-529. DOI: 10.3233/JAD-132483

[33] Kraft AD, Lee JM, Johnson DA, Kan YW, Johnson JA. Neuronal sensitivity to kainic acid is dependent on the NRF2-mediated actions of the antioxidant response element. Journal of Neurochemistry. 2006;**98**:1852-1865. DOI: 10.1111/j.1471-4159.2006.04001.x

[34] Joshi G, Gan KA, Johnson DA, Johnson JA. Increased Alzheimer's disease-like pathology in the APP/PS1ΔE9 mouse model lacking NRF2 through modulation of autophagy. Neurobiology of Aging. 2015;**36**:664-679. DOI: 10.1016/j.neurobiolaging.2014.07.030

[35] Beal MF. Therapeutic approaches to mitochondrial dysfunction in Parkinson's disease. Parkinsonism & Related Disorders. 2009;**15**(Suppl 3):S189-S194. DOI: 10.1016/s1353-8020(09)70767-5

[36] Spillantini MG, Schmidt ML, Lee VM, Trojanowski JQ, Jakes R, Goedert M. Alpha-synuclein in Lewy bodies. Nature. 1997;**388**:839-840. DOI: 10.1038/42166

[37] Irizarry MC, Growdon W, Gomez-Isla T, Newell K, George JM, Clayton DF, et al. Nigral and cortical Lewy bodies and dystrophic nigral neurites in Parkinson's disease and cortical Lewy body disease contain alpha-synuclein immunoreactivity. Journal of Neuropathology and Experimental Neurology. 1998;**57**:334-337. DOI: 10.1097/00005072-199804000-00002

[38] Spillantini MG, Crowther RA, Jakes R, Hasegawa M, Goedert M. Alpha-synuclein in filamentous inclusions of Lewy bodies from Parkinson's disease and dementia with lewy bodies. Proceedings of the National Academy of Sciences of the United States of America. 1998;**95**:6469-6473. DOI: 10.1073/pnas.95.11.6469

[39] Samii A, Nutt JG, Ransom BR. Parkinson's disease. Lancet. 2004;**363**:1783-1793. DOI: 10.1016/S0140-6736(04)16305-8

[40] Lesage S, Brice A. Parkinson's disease: From monogenic forms to genetic susceptibility factors. Human Molecular Genetics. 2009;**18**:R48-R59. DOI: 10.1093/hmg/ddp012

[41] van Muiswinkel FL, de Vos RA, Bol JG, Andringa G, Jansen Steur EN, Ross D, et al. Expression of NAD(P) H:Quinone oxidoreductase in the normal and Parkinsonian substantia nigra. Neurobiology of Aging. 2004;**25**:1253-1262. DOI: 10.1016/j.neurobiolaging.2003.12.014

[42] Castellani R, Smith MA, Richey PL, Perry G. Glycoxidation and oxidative stress in Parkinson disease and diffuse Lewy body disease. Brain Research. 1996;**737**:195-200. DOI: 10.1016/0006-8993(96)00890-9

[43] Dexter DT, Holley AE, Flitter WD, Slater TF, Wells FR, Daniel SE, et al. Increased levels of lipid hydroperoxides in the parkinsonian substantia nigra: An HPLC and ESR study. Movement Disorders. 1994;**9**:92-97. DOI: 10.1002/mds.870090113

[44] Alam ZI, Daniel SE, Lees AJ, Marsden DC, Jenner P, Halliwell B. A generalised increase in protein carbonyls in the brain in Parkinson's but not incidental Lewy body disease. Journal of Neurochemistry. 1997;**69**:1326-1329. DOI: 10.1046/j.1471-4159.1997.69031326.x

[45] Perry TL, Yong VW. Idiopathic Parkinson's disease, progressive supranuclear palsy and glutathione metabolism in the substantia nigra of patients. Neuroscience Letters. 1986;**67**:269-274. DOI: 10.1016/0304-3940(86)90355-7

[46] Zeevalk GD, Razmpour R, Bernard LP. Glutathione and Parkinson's disease: Is this the elephant in the room? Biomedicine & Pharmacotherapy. 2008;**62**:236-249. DOI: 10.1016/j.biopha.2007.06.005

[47] Imaizumi Y, Okada Y, Akamatsu W, Koike M, Kuzumaki N, Hayakawa H, et al. Mitochondrial dysfunction associated with increased oxidative stress and alpha-synuclein accumulation in PARK2 iPSC-derived neurons and postmortem brain tissue. Molecular Brain. 2012;**5**:35. DOI: 10.1186/1756-6606-5-35

[48] Rowland LP, Shneider NA. Amyotrophic lateral sclerosis. The New England Journal of Medicine. 2001;**344**:1688-1700. DOI: 10.1056/NEJM200105313442207

[49] Murayama S, Ookawa Y, Mori H, Nakano I, Ihara Y, Kuzuhara S, et al. Immunocytochemical and ultrastructural study of Lewy body-like hyaline inclusions in familial amyotrophic lateral sclerosis. Acta Neuropathologica. 1989;**78**:143-152. DOI: 10.1007/BF00691035

[50] Mizuno Y, Amari M, Takatama M, Aizawa H, Mihara B, Okamoto K. Immunoreactivities of p62, an ubiqutin-binding protein, in the spinal anterior horn cells of patients with amyotrophic lateral sclerosis. Journal of the Neurological Sciences. 2006;**249**:13-18. DOI: 10.1016/j.jns.2006.05.060

[51] Tan CF, Eguchi H, Tagawa A, Onodera O, Iwasaki T, Tsujino A, et al. TDP-43 immunoreactivity in neuronal inclusions in familial amyotrophic lateral sclerosis with or without SOD1 gene mutation. Acta Neuropathologica. 2007;**113**:535-542. DOI: 10.1007/s00401-007-0206-9

[52] Vance C, Rogelj B, Hortobagyi T, De Vos KJ, Nishimura AL, Sreedharan J, et al. Mutations in FUS, an RNA processing protein, cause familial amyotrophic lateral sclerosis type 6. Science. 2009;**323**:1208-1211. DOI: 10.1126/science.1165942

[53] Rosen DR, Siddique T, Patterson D, Figlewicz DA, Sapp P, Hentati A, et al. Mutations in Cu/Zn superoxide

dismutase gene are associated with familial amyotrophic lateral sclerosis. Nature. 1993;**362**:59-62. DOI: 10.1038/362059a0

[54] Gurney ME, Pu H, Chiu AY, Dal Canto MC, Polchow CY, Alexander DD, et al. Motor neuron degeneration in mice that express a human Cu,Zn superoxide dismutase mutation. Science. 1994;**264**:1772-1775. DOI: 10.1126/science.8209258

[55] Howland DS, Liu J, She Y, Goad B, Maragakis NJ, Kim B, et al. Focal loss of the glutamate transporter EAAT2 in a transgenic rat model of SOD1 mutant-mediated amyotrophic lateral sclerosis (ALS). Proceedings of the National Academy of Sciences of the United States of America. 2002;**99**:1604-1609. DOI: 10.1073/pnas.032539299

[56] Clement AM, Nguyen MD, Roberts EA, Garcia ML, Boillee S, Rule M, et al. Wild-type nonneuronal cells extend survival of SOD1 mutant motor neurons in ALS mice. Science. 2003;**302**:113-117. DOI: 10.1126/science.1086071

[57] Marangi G, Traynor BJ. Genetic causes of amyotrophic lateral sclerosis: New genetic analysis methodologies entailing new opportunities and challenges. Brain Research. 2015;**1607**:75-93. DOI: 10.1016/j.brainres.2015.02.003

[58] Fecto F, Yan J, Vemula SP, Liu E, Yang Y, Chen W, et al. SQSTM1 mutations in familial and sporadic amyotrophic lateral sclerosis. Archives of Neurology. 2011;**68**:1440-1446. DOI: 10.1001/archneurol.2011.250

[59] Sarlette A, Krampfl K, Grothe C, Neuhoff N, Dengler R, Petri S. Nuclear erythroid 2-related factor 2-antioxidative response element signaling pathway in motor cortex and spinal cord in amyotrophic lateral sclerosis. Journal of Neuropathology and Experimental Neurology. 2008;**67**:1055-1062. DOI: 10.1097/NEN.0b013e31818b4d40

[60] The Huntington's Disease Collaborative Research Group. A novel gene containing a trinucleotide repeat that is expanded and unstable on Huntington's disease chromosomes. Cell. 1993;**72**:971-983. DOI: 10.1016/0092-8674(93)90585-E

[61] Bates G. Huntingtin aggregation and toxicity in Huntington's disease. Lancet. 2003;**361**:1642-1644. DOI: 10.1016/S0140-6736(03)13348-4

[62] Browne SE, Beal MF. Oxidative damage in Huntington's disease pathogenesis. Antioxidants & Redox Signaling. 2006;**8**:2061-2073. DOI: 10.1089/ars.2006.8.2061

[63] Johri A, Chandra A, Beal MF. PGC-1alpha, mitochondrial dysfunction, and Huntington's disease. Free Radical Biology & Medicine. 2013;**62**:37-46. DOI: 10.1016/j.freeradbiomed.2012.09.034

[64] Reddy PH. Increased mitochondrial fission and neuronal dysfunction in Huntington's disease: Implications for molecular inhibitors of excessive mitochondrial fission. Drug Discovery Today. 2014;**19**:951-955. DOI: 10.1016/j.drudis.2014.01.021

[65] Browne SE, Ferrante RJ, Beal MF. Oxidative stress in Huntington's disease. Brain Pathology. 1999;**9**:147-163. DOI: 10.1111/j.1750-3639.1999.tb00215.x

[66] National Multiple Sclerosis Society. Who Gets MS?. 2015. Available from: http://www.nationalmssociety.org/What-is-MS/Who-Gets-MS [Accessed: February 15, 2024]

[67] Dhib-Jalbut S. Pathogenesis of myelin/oligodendrocyte damage in multiple sclerosis. Neurology. 2007;**68**:S13-S21. discussion S43-54. DOI: 10.1212/01.wnl.0000259402.47974.44

[68] Hofstetter H, Gold R, Hartung HP. Th17 cells in MS and experimental autoimmune encephalomyelitis. International MS Journal. 2009;**16**:12-18. DOI: 10.1177/1352458508101709

[69] Chiurchiu V. Novel targets in multiple sclerosis: To oxidative stress and beyond. Current Topics in Medicinal Chemistry. 2014;**14**:2590-2599. DOI: 10.2174/156802661 4666141029125739

[70] van Horssen J, Schreibelt G, Bo L, Montagne L, Drukarch B, van Muiswinkel FL, et al. NAD(P)H:quinone oxidoreductase 1 expression in multiple sclerosis lesions. Free Radical Biology & Medicine. 2006;**41**:311-317. DOI: 10.1016/j.freeradbiomed.2006.04.028

[71] van Horssen J, Schreibelt G, Drexhage J, Hazes T, Dijkstra CD, van der Valk P, et al. Severe oxidative damage in multiple sclerosis lesions coincides with enhanced antioxidant enzyme expression. Free Radical Biology & Medicine. 2008;**45**:1729-1737. DOI: 10.1016/j.freeradbiomed.2008.09.036

[72] van Horssen J, Drexhage JA, Flor T, Gerritsen W, van der Valk P, de Vries HE. NRF2 and DJ1 are consistently upregulated in inflammatory multiple sclerosis lesions. Free Radical Biology & Medicine. 2010;**49**:1283-1289. DOI: 10.1016/j.freeradbiomed.2010.07.032

[73] Prasad KN. Simultaneous activation of Nrf2 and elevation of antioxidant compounds for reducing oxidative stress and chronic inflammation in human Alzheimer's disease. Mechanisms of Ageing and Development. 2016;**153**:41-47. DOI: 10.1016/j.mad.2016.01.002

[74] Rojo AI, McBean G, Cindric M, Egea J, Lopez MG, Rada P, et al. Redox control of microglial function: Molecular mechanisms and functional significance. Antioxidants & Redox Signaling. 2014;**21**:1766-1801. DOI: 10.1089/ars.2013.5748

[75] Lawson LJ, Perry VH, Dri P, Gordon S. Heterogeneity in the distribution and morphology of microglia in the normal adult mouse brain. Neuroscience. 1990;**39**:151-170. DOI: 10.1016/0306-4522(90)90229-W

[76] Lull ME, Block ML. Microglial activation and chronic neurodegeneration. Neurotherapeutics. 2010;7:354-365. DOI: 10.1016/j.nurt.2010.05.014

[77] Innamorato NG, Rojo AI, García-Yagüe AJ, Yamamoto M, de Ceballos ML, Cuadrado A. The transcription factor NRF2 is a therapeutic target against brain inflammation. Journal of Immunology. 2008;**181**(1):680-689. DOI: 10.4049/jimmunol.181.1.680

[78] Wang X, de Rivero Vaccari JP, Wang H, et al. Activation of the nuclear factor E2-related factor 2/antioxidant response element pathway is neuroprotective after spinal cord injury. Journal of Neurotrauma. 2012;**29**(5):936-945. DOI: 10.1089/neu.2011.1922

[79] Rojo AI, Innamorato NG, Martín-Moreno AM, De Ceballos ML, Yamamoto M, Cuadrado A. NRF2 regulates microglial dynamics and neuroinflammation in experimental Parkinson's disease. Glia.

2010;**58**(5):588-598. DOI: 10.1002/glia.20947

[80] Chen PC, Vargas MR, Pani AK, et al. NRF2-mediated neuroprotection in the MPTP mouse model of Parkinson's disease: Critical role for the astrocyte. Proceedings of the National Academy of Sciences of the United States of America. 2009;**106**(8):2933-2938. DOI: 10.1073/pnas.0813361106

[81] Lastres-Becker I, Innamorato NG, Jaworski T, et al. Fractalkine activates NRF2/NFE2L2 and heme oxygenase 1 to restrain tauopathy-induced microgliosis. Brain. 2014;**137**(Pt 1):78-91. DOI: 10.1093/brain/awt323

[82] Li E, Noda M, Doi Y, et al. The neuroprotective effects of milk fat globule-EGF factor 8 against oligomeric amyloid β toxicity. Journal of Neuroinflammation. 2012;**9**:148. DOI: 10.1186/1742-2094-9-148

[83] Dinkova-Kostova AT, Rumen VK, Kazantsev AG. The role of Nrf2 signaling in counteracting neurodegenerative diseases. The FEBS Journal. 2018;**285**:3576-3590. DOI: 10.1111/febs.14379

[84] LoGerfo A, Chico L, Borgia L, et al. Lack of association between nuclear factor erythroid-derived 2-like 2 promoter gene polymorphisms and oxidative stress biomarkers in amyotrophic lateral sclerosis patients. Oxidative Medicine and Cellular Longevity. 2014;**2014**:432626. DOI: 10.1155/2014/432626

[85] von Otter M, Landgren S, Nilsson S, et al. Association of NRF2-encoding NFE2L2 haplotypes with Parkinson's disease. BMC Medical Genetics. 2010;**11**:36. DOI: 10.1186/1471-2350-11-36

[86] von Otter M, Bergstrom P, Quattrone A, De Marco EV, Annesi G, Soderkvist P, et al. Genetic associations of NRF2-encoding NFE2L2 variants with Parkinson's disease—A multicenter study. BMC Medical Genetics. 2014;**15**:131. DOI: 10.1186/s12881-014-0131-4

[87] Chen YC, Wu YR, Wu YC, Lee-Chen GJ, Chen CM. Genetic analysis of NFE2L2 promoter variation in Taiwanese Parkinson's disease. Parkinsonism & Related Disorders. 2013;**19**(2):247-250. DOI: 10.1016/j.parkreldis.2012.10.018

[88] Rojo AI, Pajares M, García-Yagüe AJ, et al. Deficiency in the transcription factor NRF2 worsens inflammatory parameters in a mouse model with combined tauopathy and amyloidopathy. Redox Biology. 2018;**18**:173-180. DOI: 10.1016/j.redox.2018.07.006

[89] Bergström P, von Otter M, Nilsson S, et al. Association of NFE2L2 and KEAP1 haplotypes with amyotrophic lateral sclerosis. Amyotrophic Lateral Sclerosis and Frontotemporal Degeneration. 2014;**15**(1-2):130-137. DOI: 10.3109/21678421.2013.839708

[90] Wang L, Lewis T, Zhang YL, et al. The identification and characterization of non-reactive inhibitor of Keap1-NRF2 interaction through HTS using a fluorescence polarization assay. In: Probe Reports from the NIH Molecular Libraries Program. Bethesda (MD): National Center for Biotechnology Information (US); 2010

[91] Hancock R, Bertrand HC, Tsujita T, et al. Peptide inhibitors of the Keap1-NRF2 protein-protein interaction. Free Radical Biology & Medicine. 2012;**52**(2):444-451. DOI: 10.1016/j.freeradbiomed.2011.10.486

[92] Zhuang C, Miao Z, Sheng C, Zhang W. Updated research and applications of small molecule inhibitors of Keap1-NRF2 protein-protein interaction: A review. Current Medicinal Chemistry. 2014;**21**(16):1861-1870. DOI: 10.2174/0929867321666140217104648

[93] de Zeeuw D, Akizawa T, Audhya P, et al. Bardoxolone methyl in type 2 diabetes and stage 4 chronic kidney disease. The New England Journal of Medicine. 2013;**369**(26):2492-2503. DOI: 10.1056/NEJMoa1306033

[94] Reisman SA, Ward KW, Klaassen CD, Meyer CJ. CDDO-9,11-dihydro-trifluoroethyl amide (CDDO-dhTFEA) induces hepatic cytoprotective genes and increases bile flow in rats. Xenobiotica. 2013;**43**(7):571-578. DOI: 10.3109/00498254.2012.750022

[95] Li B, Abdalrahman A, Lai Y, Janicki JS, Ward KW, Meyer CJ, et al. Dihydro-CDDO-trifluoroethyl amide suppresses inflammatory responses in macrophages via activation of NRF2. Biochemical and Biophysical Research Communications. 2014;**444**:555-561. DOI: 10.1016/J.BBRC.2014.01.101

[96] Reisman SA, Lee CY, Meyer CJ, Proksch JW, Ward KW. Topical application of the synthetic triterpenoid RTA 408 activates NRF2 and induces cytoprotective genes in rat skin. Archives of Dermatological Research. 2014;**306**(5):447-454. DOI: 10.1007/s00403-013-1433-7

[97] Probst BL, Trevino I, McCauley L, Bumeister R, Dulubova I, Wigley WC, et al. A novel synthetic triterpenoid with broad anticancer and anti-inflammatory activity. PLoS ONE. 2015;**10**:e0122942. DOI: 10.1371/journal.pone.0122942